Honors Physics Essentials

An APlusPhysics Guide

Dan Fullerton

Physics Teacher
Irondequoit High School

Adjunct Professor
Microelectronic Engineering
Rochester Institute of Technology

Dedication

To Andrea, you make my heart smile.
To Piglet, for your inspiration and love.
To my parents, for their amazing support.
And to my grandparents, for their steadfast dedication.
To UB, AP and AD, who always encouraged me to dream big.
To WB and KF, for allowing me to take risks, make mistakes, and learn.
And finally to my students, for making my job the best job in the world.

Credits

Thanks To:
Mike Powlin, Jeff Guercio, Jeff Yap, Tom Schulte, Mike Jackson, Geoffrey Clarion,
Byron Philhour, Marcie Shea, Joe Kunz, Santosh Kurinec, and Doni Parnell
for technical input, review, and guidance.

Silly Beagle Productions
656 Maris Run
Webster, NY 14580
Internet: www.SillyBeagle.com
E-Mail: info@SillyBeagle.com

Cover Design: Dan Fullerton
Interior Illustrations by Dan Fullerton, Jupiterimages and NASA unless otherwise noted
All images and illustrations ©2011 Jupiterimages Corporation and Dan Fullerton
Edited by Jeff Guercio

Sales and Ordering Information
http://www.aplusphysics.com/honors
Sales@SillyBeagle.com
Volume discounts available.
E-book editions available.

Printed in the United States of America
ISBN: 978-0-9835633-1-0

1 2 3 4 5 6 7 8 9 0 9 8 7 6M

Silly Beagle Productions

Welcome to Honors Physics Essentials, an APlusPhysics Guide. From mechanics, fluids, thermodynamics, and electricity to waves, optics, modern physics, and semiconductors, this book is your essential physics resource for use as a standalone guide; companion to classroom texts; or as a review book for your high school Honors Physics course. Further, because this book provides an overview of so many areas of physics, it's a great tool in preparing for standardized exams such as the SAT Subject Test in Physics, NY Regents Physics Exam, algebra-based AP Physics, and IB Physics.

What sets this book apart from the other review books?

1. It reviews the essential concepts and understandings required for success in a majority of algebra-based high school physics courses.
2. It includes more than 500 sample questions with full solutions, integrated into the chapters immediately following the material being covered, so you can test your understanding.
3. It is supplemented by the free APlusPhysics.com web site, which includes:
   a. Videos and tutorials on key physics concepts
   b. Interactive practice quizzes
   c. Discussion and homework help forums supported by the author and fellow readers
   d. Student blogs to share challenges, successes, hints and tricks
   e. Projects and activities designed to improve your understanding of essential physics concepts in a fun and engaging manner
   f. Latest and greatest physics news

Just remember, physics is fun! It's an exciting course, and with a little preparation and this book, you can transform your quest for essential physics comprehension from a stressful chore into an enjoyable and, yes, FUN, opportunity for success.

How to Use This Book

This book is arranged by topic, with sample problems and solutions integrated right in the text. Actively explore each chapter. Cover up the in-text solutions with an index card, get out a pencil, and try to solve the sample problems yourself as you go, before looking at the answer. If you're stuck, don't stress! Post your problem on the APlusPhysics website (http://aplusphysics.com) and get help from other students, teachers, and subject matter experts (including the author of this book!) Once you feel confident with the subject matter, test yourself and see how you perform. Review areas of difficulty, then try again and watch your understanding improve!

Table of Contents

Chapter 1: Introduction

"There is a theory which states that if ever anybody discovers exactly what the Universe is for and why it is here, it will instantly disappear and be replaced by something even more bizarre and inexplicable.

There is another theory which states that this has already happened."

— *Douglas Adams*

Objectives

Explore the scope of physics and review pre-requisite skills for success.

1. Recognize the questions of physics.
2. List several disciplines within the study of physics.
3. Define matter, mass, work and energy.
4. Express answers correctly with respect to significant figures.
5. Use scientific notation to express physical values efficiently.
6. Convert and estimate SI units.
7. Differentiate between scalar and vector quantities.
8. Use scaled diagrams to represent and manipulate vectors.
9. Determine x- and y-components of two-dimensional vectors.
10. Determine the angle of a vector given its components.

What Is Physics?

Physics is many things to many different people. If you look up physics in the dictionary, you'll probably learn physics has to do with matter, energy, and their interactions. But what does that really mean? What is matter? What is energy? How do they interact? And most importantly, why do we care?

Physics, in many ways, is the answer to the favorite question of most 2-year-olds: "Why?" What comes after the why really doesn't matter. If it's a "why" question, chances are it's answered by physics. Why is the sky blue? Why does the wind blow? Why do we fall down? Why does my teacher smell funny? Why do airplanes fly? Why do the stars shine? Why do I have to eat my vegetables? The answer to all these questions, and many more, ultimately reside in the realm of physics.

Matter

If physics is the study of matter, then we probably ought to define matter. **Matter**, in scientific terms, is anything that has mass and takes up space. In short, then, matter is anything you can touch – from objects smaller than electrons to stars hundreds of times larger than the sun. From this perspective, physics is the mother of all science. Astronomy to zoology, all other branches of science are subsets of physics, or specializations inside the larger discipline of physics.

So what's mass? **Mass** is, in simple terms, the amount of "stuff" an object is made up of. But of course, there's more to the story. Mass is split into two types... **inertial mass** and **gravitational mass**. Inertial mass is an object's resistance to being accelerated by a force. More massive objects accelerate less than smaller objects given an identical force. Gravitational mass, on the other hand, relates to the amount of gravitational force experienced by an object. Objects with larger gravitational mass experience a larger gravitational force.

Confusing? Don't worry! As it turns out, in all practicality, inertial mass and gravitational mass have always been equal for any object measured, even if it's not immediately obvious why this is the case (although with an advanced study of Einstein's Theory of General Relativity you can predict this outcome).

1.1 Q: On the surface of Earth, a spacecraft has a mass of 2.00×10^4 kg. What is the mass of the spacecraft at a distance of one Earth radius above Earth's surface?

(1) 5.00×10^3 kg

(2) 2.00×10^4 kg

(3) 4.90×10^4 kg

(4) 1.96×10^5 kg

1.1 A: (2) 2.00×10^4 kg. Mass is constant, therefore the spacecraft's mass at a distance of one Earth radius above Earth's surface is 2.00×10^4 kg.

Energy

If it's not matter, what's left? Why, energy, of course. As energy is such an everyday term that encompasses so many areas, an accurate definition can be quite elusive. Physics texts oftentimes define **energy** as the ability or capacity to do work. It's a nice, succinct definition, but leads to another question – what is work? **Work** can also be defined many ways, but a general definition starts with the process of moving an object. If you put these two definitions together, you can vaguely define energy as the ability or capacity to move an object.

Mass – Energy Equivalence

So far, the definition of physics boils down to the study of matter, energy, and their interactions. Around the turn of the 20th century, however, several physicists began proposing a strong relationship between matter and energy. Albert Einstein, in 1905, formalized this with his famous formula $E=mc^2$, which states that the mass of an object, a key characteristic of matter, is really a measure of its energy. This discovery has paved the way for tremendous innovation ranging from nuclear power plants to atomic weapons to particle colliders performing research on the origins of the universe. Ultimately, if traced back to its origin, the source of all energy on earth is the conversion of mass to energy!

Scope of Honors Physics

Physics, in some sense, can therefore by defined as the study of just about everything. Try to think of something that isn't physics – go on, I dare you! Not so easy, is it? Even the more ambiguous topics can, in some sense, be categorized as physics. A Shakespearean sonnet? A sonnet is typically read from a manuscript (matter), and sensed by the conversion of light (energy) alternately reflected and absorbed from a substrate, focused by a lens in the eye, and converted to chemical and electrical signals by photoreceptors on the retina, and then transferred as electrical and chemical signals along the neural pathways to the brain. In short, just about everything is physics from a certain perspective.

As this is an introductory book in physics, it's important to limit the scope of study as a foundational understanding of the world is developed. Beginning with a study of Newtonian Mechanics, which explores moving objects and their interactions, you can then talk about the mechanics of fluids, and continue on by talking about thermal energy and its transfer. From there, you can transition into electricity and magnetism. Next, you'll briefly explore how the application of physics and engineering principles to semiconductors has changed (and continues to change) the world. Then, you'll explore the transfer of energy using waves. Next, using what you've learned about physical waves, you can expand into electromagnetic waves and optics. Finally, you can conclude by looking at matter again, this time at the nuclear level, using your background in mechanics, electricity and magnetism, waves, and optics to build a deeper understanding of the building blocks of existence in the Modern Physics chapter.

Join me in taking the first steps into a better understanding of the universe we live in.

Significant Figures

Significant Figures (or sig figs, for short) represent a manner of showing which digits in a number are known to some level of certainty. But how do you know which digits are significant? There are some rules to help with this. If you start with a number in scientific notation:

- All non-zero digits are significant.
- All digits between non-zero digits are significant.
- Zeroes to the left of significant digits are not significant.
- Zeroes to the right of significant digits are significant.

When you make a measurement in physics, you want to write what you measured using significant figures. To do this, write down as many digits as you are absolutely certain of, then take a shot at one more digit as accurately as you can. These are your significant figures.

1.2 Q: How many significant figures are in the value 43.74 km?

1.2 A: 4 (four non-zero digits)

1.3 Q: How many significant figures are in the value 4302.5 g?

1.3 A: 5 (All non-zero digits are significant and digits between non-zero digits are significant.)

1.4 Q: How many significant figures are in the value 0.0083s?

1.4 A: 2 (All non-zero digits are significant. Zeroes to the left of significant digits are not significant.)

1.5 Q: How many significant figures are in the value 1.200×10^3 kg?

1.5 A: 4 (Zeroes to the right of significant digits are significant.)

Scientific Notation

Although physics and mathematics aren't the same thing, they are in many ways closely related. Just like English is the language of this content, mathematics is the language of physics. A solid understanding of a few simple math concepts will allow you to communicate and describe the physical world both efficiently and accurately.

Because measurements of the physical world vary so tremendously in size (imagine trying to describe the distance across the United States in units of hair thicknesses), physicists often times use what is known as **scientific notation** to represent very large and very small numbers. These very large and very small numbers would become quite cumbersome to write out repeatedly. Imagine writing 4,000,000,000,000 over and over again. Your hand would get tired and your pen would rapidly run out of ink! Instead, it's much easier to write this number as 4×10^{12}. Or on the smaller scale, the thickness of the insulating layer (known as a gate dielectric) in the integrated circuits that power computers and other electronics can be less than 0.000000001 m. It's easy to lose track of how many zeros you have to deal with, so scientists instead would write this number as 1×10^{-9} m. See how much simpler life can be with scientific notation?

Scientific notation follows these simple rules. Start by showing all the significant figures in the number you're describing, with the decimal point after the first significant digit. Then, show your number being multiplied by 10 to the appropriate power in order to give you the correct value.

It sounds more complicated than it is. Let's say, for instance, you want to show the number 300,000,000 in scientific notation (a very useful number in physics), and let's assume you know this value to three significant digits. You would start by writing the three significant digits, with the decimal point after the first digit, as "3.00". Now, you need to multiply this number by 10 to some power in order to get back to the original value. In this case, you multiply 3.00 by 10^8, for an answer of 3.00×10^8. Interestingly, the power you raise the 10 to is exactly equal to the number of digits you moved the decimal to the left as you converted from standard to scientific notation. Similarly, if you start in scientific notation, to convert to standard notation, all you have to do is remove the 10^8 power by moving the decimal point eight digits to the right. Presto, you're an expert in scientific notation!

But, what do you do if the number is much smaller than one? Same basic idea... let's assume you're dealing with the approximate radius of an electron, which is 0.00000000000000282 m. It's easy to see how unwieldy this could become. You can write this in scientific notation by writing out three significant digits, with the decimal point after the first digit, as "2.82." Again, you multiply this number by some power of 10 in order to get back to the original value. Because your value is less than 1, you need to use negative powers of 10. If you raise 10 to the power -15, specifically, you get a final value of 2.82×10^{-15} m. In essence, for every digit you moved the decimal place, you add another power of 10. And if you start with scientific notation, all you do is move the decimal place left one digit for every negative power of 10.

1.6 Q: Express the number 0.000470 in scientific notation.
1.6 A: 4.70×10^{-4}

1.7 Q: Express the number 2,870,000 in scientific notation.
1.7 A: 2.87×10^6

1.8 Q: Expand the number 9.56×10^{-3}.
1.8 A: 0.00956

1.9 Q: Expand the number 1.11×10^7.
1.9 A: 11,100,000

Metric System

Physics involves the study, prediction, and analysis of real-world phenomena. To communicate data accurately, you must set specific standards for basic measurements. The physics community has standarized on what is known as the **Système International** (SI), which defines seven baseline measurements and their standard units, forming the foundation of what is called the metric system of measurement. The SI system is oftentimes referred to as the **mks system**, as the three most common measurement units are meters, kilograms, and seconds, which will be the focus for the majority of this course. The fourth SI base unit you'll use in this course, the ampere, will be introduced in the current electricity section.

The base unit of length in the metric system, the meter, is roughly equivalent to the English yard. For smaller measurements, the meter is divided up into 100 parts, known as centimeters, and each centimeter is made up of 10 millimeters. For larger measurements, the meter is grouped into larger units of 1000 meters, known as a kilometer. The length of a baseball bat is approximately one meter, the radius of a U.S. quarter is approximately a centimeter, and the diameter of the metal in a wire paperclip is roughly one millimeter.

The base unit of mass, the kilogram, is roughly equivalent to two U.S. pounds. A cube of water 10 cm x 10 cm x 10 cm has a mass of 1 kilogram. Kilograms can also be broken up into larger and smaller units, with commonly used measurements of grams (1/1000th of a kilogram) and milligrams (1/1000th of a gram). The mass of a textbook is approximately 2 to 3 kilograms, the mass of a baseball is approximately 145 grams, and the mass of a mosquito is 1 to 2 grams.

The base unit of time, the second, is likely already familiar. Time can also be broken up into smaller units such as milliseconds (10^{-3} seconds), microseconds (10^{-6} seconds), and nanoseconds (10^{-9} seconds), or grouped into larger units such as minutes (60 seconds), hours (60 minutes), days (24 hours), and years (365.25 days).

The metric system is based on powers of 10, allowing for easy conversion from one unit to another. A chart showing the meaning of commonly used metric prefixes and their notations can be extremely valuable in performing unit conversions.

Prefixes for Powers of 10

Prefix	Symbol	Notation
tera	T	10^{12}
giga	G	10^{9}
mega	M	10^{6}
kilo	k	10^{3}
deci	d	10^{-1}
centi	c	10^{-2}
milli	m	10^{-3}
micro	μ	10^{-6}
nano	n	10^{-9}
pico	p	10^{-12}

Converting from one unit to another can be easily accomplished if you use the following procedure.

1. Write your initial measurement with units as a fraction over 1.
2. Multiply your initial fraction by a second fraction, with a numerator (top number) having the units you want to convert to, and the denominator (bottom number) having the units of your initial measurement.
3. For any units on the top right-hand side with a prefix, determine the value for that prefix. Write that prefix in the right-hand denominator. If there is no prefix, use 1.
4. For any units on the right-hand denominator with a prefix, write the value for that prefix in the right-hand numerator. If there is no prefix, use 1.
5. Multiply through the problem, taking care to accurately record units. You should be left with a final answer in the desired units.

Let's take a look at a sample unit conversion:

1.10 Q: Convert 23 millimeters (mm) to meters (m).

1.10 A: Step 1. $\dfrac{23mm}{1}$

Step 2. $\dfrac{23mm}{1} \times \dfrac{m}{mm}$

Step 3. $\dfrac{23mm}{1} \times \dfrac{m}{1mm}$

Step 4. $\dfrac{23mm}{1} \times \dfrac{10^{-3}\,m}{1mm}$

Step 5. $\dfrac{23mm}{1} \times \dfrac{10^{-3}\,m}{1mm} = 2.3 \times 10^{-2}\,m$

Now, try some on your own!

1.11 Q: Convert 2.67×10⁻⁴ m to mm.

1.11 A: $\dfrac{2.67 \times 10^{-4}\,m}{1} \times \dfrac{1mm}{10^{-3}\,m} = 0.267\,mm$

1.12 Q: Convert 14 kg to mg.

1.12 A: $\dfrac{14kg}{1} \times \dfrac{10^{3}\,mg}{10^{-3}\,kg} = 14 \times 10^{6}\,mg$

1.13 Q: Convert 3,470,000 μs to s.

1.13 A: $\dfrac{3,470,000\mu s}{1} \times \dfrac{10^{-6}\,s}{1\mu s} = 3.47\,s$

1.14 Q: Convert 64 GB to KB and express in scientific notation.

1.14 A: $\dfrac{64GB}{1} \times \dfrac{10^{9}\,KB}{10^{3}\,GB} = 64,000,000\,KB = 6.4 \times 10^{7}\,KB$

Accuracy and Precision

When making measurements of physical quantities, how close the measurement is to the actual value is known as the **accuracy** of the measurement. **Precision**, on the other hand, is the repeatability of a measurement. A common analogy involves an archer shooting arrows at the target. The bullseye of the target represents the actual value of the measurement.

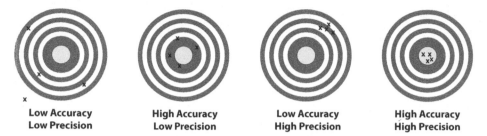

| Low Accuracy | High Accuracy | Low Accuracy | High Accuracy |
| Low Precision | Low Precision | High Precision | High Precision |

Ideally, measurements in physics should be both accurate and precise.

Algebra and Trigonometry

Just as you find the English language a convenient tool for conveying your thoughts to others, you need a convenient language for conveying your understanding of the world around you in order to understand its behavior. The language most commonly (and conveniently) used to describe the natural world is mathematics. Therefore, to understand physics, you need to be fluent in the mathematics of the topics you'll study in this book... specifically basic algebra and trigonometry.

Now don't you fret or frown. You need only the most basic of algebra and trigonometry knowledge in order to successfully solve a wide range of physics problems.

A vast majority of problems requiring algebra can be solved using the same problem solving strategy. First, analyze the problem and write down what you know, what you need to find, and make a picture or diagram to better understand the problem if it makes sense. Then, start your solution by searching for a path that will lead you from your givens to your finds. Once you've determined an appropriate pathway (and there may be more than one), solve your problem algebraically for your answer. Finally, as your last steps, substitute in any values with units into your final equation, and solve for your answer, with units.

The use of trigonometry, the study of right triangles, can be distilled down to the definitions of the three basic trigonometric functions. When you know the length of two sides of a right triangle, or the length of one side and a non-right angle, you can solve for all the angles and sides of the triangle.If you can use the definitions of the sine, cosine, and tangent, you'll be fine in physics.

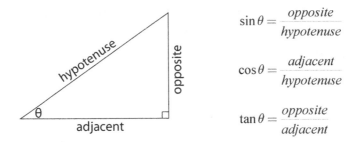

$$\sin\theta = \frac{opposite}{hypotenuse}$$

$$\cos\theta = \frac{adjacent}{hypotenuse}$$

$$\tan\theta = \frac{opposite}{adjacent}$$

Of course, if you need to solve for the angles themselves, you can use the inverse trigonometric functions.

$$\theta = \sin^{-1}\left(\frac{opposite}{hypotenuse}\right) = \cos^{-1}\left(\frac{adjacent}{hypotenuse}\right) = \tan^{-1}\left(\frac{opposite}{adjacent}\right)$$

1.15 Q: A car travels from the airport 14 miles east and 7 miles north to its destination. What direction should a helicopter fly from the airport to reach the same destination, traveling in a straight line?

1.15 A:

$$\theta = \tan^{-1}\left(\frac{opposite}{adjacent}\right)$$

$$\theta = \tan^{-1}\left(\frac{7\ miles}{14\ miles}\right) = 26.6°$$

Vectors and Scalars

Quantities in physics are used to represent real-world measurements, and therefore physicists use these quantities as tools to better understand the world. In examining these quantities, there are times when just a number, with a unit, can completely describe a situation. These numbers, which have just a **magnitude**, or size, are known as **scalars**. Examples of scalars include quantities such as temperature, mass, and time. At other times, a quantity is more descriptive if it also includes a direction. These quantities which have both a magnitude and direction are known as **vectors**. Vector quantities you may be familiar with include force, velocity, and acceleration.

Most students will be familiar with scalars, but to many, vectors may be a new and confusing concept. By learning just a few rules for dealing with vectors, you'll find that they can be a powerful tool for problem solving.

Vectors are often represented as arrows, with the length of the arrow indicating the magnitude of the quantity, and the direction of the arrow indicating the direction of the vector. In the figure below, vector B has a magnitude greater than that of vector A even though vectors A and B point in the same direction. It's also important to know that vectors can be moved anywhere in space. The positions of A and B could be reversed, and the individual vectors would retain their values of magnitude and direction.

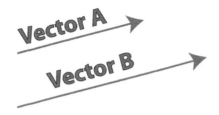

To add vectors A and B below, all you have to do is line them up so that the tip of the first vector touches the tail of the second vector. Then, to find the sum of the vectors, known as the **resultant**, draw a straight line from the start of the first vector to the end of the last vector. This method works with any number of vectors.

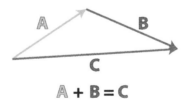

$$A + B = C$$

So how do you subtract two vectors? Try subtracting B from A. You can start by rewriting the expression A - B as A + -B. Now it becomes an addition problem, you just have to figure out how to express –B. This is easier than it sounds. To find the opposite of a vector, just point the vector in the opposite direction. Therefore, you can use what we already know about the addition of vectors to find the resultant of A-B.

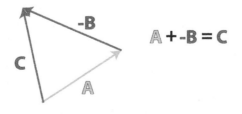

$$A + -B = C$$

Components of Vectors

You'll learn more about vectors as you go, but before moving on, there are a few basic skills to master. Vectors at angles can be challenging to deal with. By transforming a vector at an angle into two vectors, one parallel to the x-axis and one parallel to the y-axis, you can greatly simplify problem solving. To break a vector up into its components, you can use the basic trig functions.

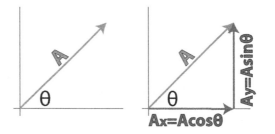

1.16 Q: The vector diagram below represents the horizontal component, F_H, and the vertical component, F_V, of a 24-Newton force acting at 35° above the horizontal. What are the magnitudes of the horizontal and vertical components?

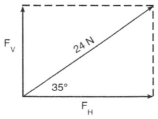

(1) F_H=3.5 N and F_V=4.9 N
(2) F_H=4.9 N and F_V=3.5 N
(3) F_H=14 N and F_V=20 N
(4) F_H=20 N and F_V=14 N

1.16 A: (4) $F_H = A_x = A\cos\theta = (24N)\cos 35° = 20N$

$F_V = A_y = A\sin\theta = (24N)\sin 35° = 14N$

1.17 Q: An airplane flies with a velocity of 750 kilometers per hour, 30° south of east. What is the magnitude of the plane's eastward velocity?

(1) 866 km/h
(2) 650 km/h
(3) 433 km/h
(4) 375 km/h

1.17 A:

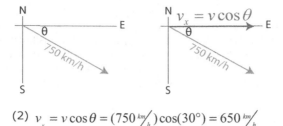

(2) $v_x = v\cos\theta = (750\ ^{km}\!/_h)\cos(30°) = 650\ ^{km}\!/_h$

1.18 Q: A soccer player kicks a ball with an initial velocity of 10 m/s at an angle of 30° above the horizontal. The magnitude of the horizontal component of the ball's velocity is

(1) 5.0 m/s
(2) 8.7 m/s
(3) 9.8 m/s
(4) 10 m/s

1.18 A:

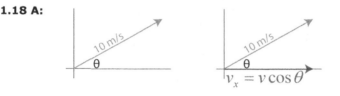

(2) $v_x = v\cos\theta = (10\ ^m\!/_s)\cos(30°) = 8.7\ ^m\!/_s$

1.19 Q: A child kicks a ball with an initial velocity of 8.5 meters per second at an angle of 35° with the horizontal, as shown. The ball has an initial vertical velocity of 4.9 meters per second. The horizontal component of the ball's initial velocity is approximately

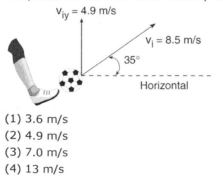

$v_{iy} = 4.9$ m/s

$v_i = 8.5$ m/s

35°

Horizontal

(1) 3.6 m/s
(2) 4.9 m/s
(3) 7.0 m/s
(4) 13 m/s

1.19 A: (3) $v_x = v\cos\theta = (8.5\ ^m\!/_s)\cos(35°) = 6.96\ ^m\!/_s$

In similar fashion, you can use the components of a vector in order to build the original vector. Graphically, if you line up the component vectors tip-to-tail, the original vector runs from the starting point of the first vector to the ending point of the last vector. To determine the magnitude of the resulting vector algebraically, just apply the Pythagorean Theorem.

1.20 Q: A motorboat, which has a speed of 5.0 meters per second in still water, is headed east as it crosses a river flowing south at 3.3 meters per second. What is the magnitude of the boat's resultant velocity with respect to the starting point?

(1) 3.3 m/s

(2) 5.0 m/s

(3) 6.0 m/s

(4) 8.3 m/s

1.20 A: (3) 6.0 m/s

The motorboat's resultant velocity is the vector sum of the motorboat's speed and the riverboat's speed.

$$a^2 + b^2 = c^2$$
$$c = \sqrt{a^2 + b^2}$$
$$c = \sqrt{\left(5\,\tfrac{m}{s}\right)^2 + \left(3.3\,\tfrac{m}{s}\right)^2}$$
$$c = 6\,\tfrac{m}{s}$$

1.21 Q: A dog walks 8.0 meters due north and then 6.0 meters due east. Determine the magnitude of the dog's total displacement.

1.21 A:
$$a^2 + b^2 = c^2$$
$$c = \sqrt{a^2 + b^2}$$
$$c = \sqrt{\left(6\,\tfrac{m}{s}\right)^2 + \left(8\,\tfrac{m}{s}\right)^2}$$
$$c = 10\,\tfrac{m}{s}$$

1.22 Q: A 5.0-newton force could have perpendicular components of

(1) 1.0 N and 4.0 N

(2) 2.0 N and 3.0 N

(3) 3.0 N and 4.0 N

(4) 5.0 N and 5.0 N

1.22 A: (3) The only answers that fit the Pythagorean Theorem are 3.0 N and 4.0 N ($3^2 + 4^2 = 5^2$)

1.23 Q: A vector makes an angle, θ, with the horizontal. The horizontal and vertical components of the vector will be equal in magnitude if angle θ is

(1) 30°

(2) 45°

(3) 60°

(4) 90°

1.23 A: (2) 45°. A_x=Acos(θ) will be equal to A_y=Asin(θ) when angle θ=45° since cos(45°)=sin(45°).

The Equilibrant Vector

The **equilibrant** of a force vector or set of force vectors is a single force vector which is exactly equal in magnitude and opposite in direction to the original vector or sum of vectors. The equilibrant, in effect, "cancels out" the original vector(s), or brings the set of vectors into equilibrium. To find an equilibrant, first find the resultant of the original vectors. The equilibrant is the opposite of the resultant you found!

1.24 Q: The diagram below represents two concurrent forces.

Which vector represents the force that will produce equilibrium with these two forces?

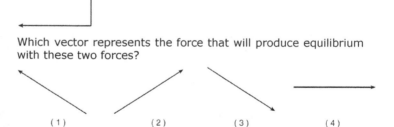

(1)   (2)   (3)   (4)

1.24 A: (3) The resultant of the two vectors would point up and to the left, therefore the equilibrant must point in the opposite direction, down and to the right.

You can find more practice problems on the APlusPhysics website at: http://www.aplusphysics.com.

Chapter 2: Kinematics

"I like physics. I think it is the best science out of all three of them, because generally it's more useful. You learn about speed and velocity and time, and that's all clever stuff."

— Tom Felton

Objectives

1. Understand the difference between position, distance and displacement, and between speed and velocity.
2. Calculate distance, displacement, speed, velocity, and acceleration.
3. Solve problems involving average speed and average velocity.
4. Construct and interpret graphs of position, velocity, and acceleration versus time.
5. Determine and interpret slopes and areas of motion graphs.
6. Determine the acceleration due to gravity near the surface of Earth.
7. Use kinematic equations to solve problems for objects moving at a constant acceleration in a straight line and in free fall.
8. Resolve a vector into perpendicular components: both graphically and algebraically.
9. Sketch the theoretical path of a projectile.
10. Recognize the independence of the vertical and horizontal motions of a projectile.
11. Solve problems involving projectile motion for projectiles fired horizontally and at an angle.
12. Solve problems involving changing frames of reference and relative velocities.

Physics is all about energy in the universe, in all its various forms. Here on Earth, the source of all energy, directly or indirectly, is the conversion of mass into energy. We receive most of this energy from the sun. Solar power, wind power, hydroelectric power, fossil fuels, all can be traced back to the sun and the conversion of mass into energy. So where do you start in your study of the universe?

Theoretically, you could start by investigating any of these types of energy. In reality, however, by starting with energy of motion (also known as **kinetic energy**), you can develop a set of analytical problem solving skills from basic principles that will serve you well as you expand into the study of other types of energy.

For an object to have kinetic energy, it must be moving. Specifically, the kinetic energy of an object is equal to one half of the object's mass multiplied by the square of its velocity.

$$KE = \tfrac{1}{2}mv^2$$

If kinetic energy is energy of motion, and energy is the ability or capacity to do work (moving an object), then you can think of kinetic energy as the ability or capacity of a moving object to move another object.

But what does it mean to be in motion? A moving object has a varying position. Its location changes as a function of time. So to understand kinetic energy, you'll need to better understand position and how position changes. This will lead into the first major unit, kinematics, from the Greek word kinein, meaning to move. Formally, kinematics is the branch of physics dealing with the description of an object's motion, leaving the study of the "why" of motion to the next major topic, dynamics.

Position, Distance and Displacement

An object's **position** refers to its location at any given point in time. Position is a vector, and its magnitude is given by the symbol **x**. If we confine our study to motion in one dimension, we can define how far an object travels from its initial position as its **distance**. Distance, as defined by physics, is a **scalar**. It has a magnitude, or size, only. The basic unit of distance is the meter (m).

2.1 Q: On a sunny afternoon, a deer walks 1300 meters east to a creek for a drink. The deer then walks 500 meters west to the berry patch for dinner, before running 300 meters west when startled by a loud raccoon. What distance did the deer travel?

2.1 A: The deer traveled 1300m + 500m + 300m, for a total distance traveled of 2100m.

Besides distance, in physics it is also helpful to know how far an object is from its starting point, or its change in position. The vector quantity **displacement** $(x-x_0)$, or Δx, describes how far an object is from its starting point, and the direction of the displacement vector points from the starting point to the finishing point. Like distance, the units of displacement are meters.

2.2 Q: A deer walks 1300 m east to a creek for a drink. The deer then walked 500 m west to the berry patch for dinner, before running 300 m west when startled by a loud raccoon. What is the deer's displacement?

2.2 A: The deer's displacement was 500m east.

2.3 Q: Which is a vector quantity?
(1) speed
(2) work
(3) mass
(4) displacement

2.3 A: (4) Displacement is a vector quantity; it has direction.

2.4 Q: A student on her way to school walks four blocks east, three blocks north, and another four blocks east, as shown in the diagram.

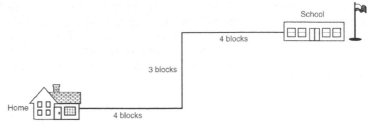

Compared to the distance she walks, the magnitude of her displacement from home to school is

(1) less

(2) greater

(3) the same

2.4 A: (1) The magnitude of displacement is always less than or equal to the distance traveled.

2.5 Q: A hiker walks 5 kilometers due north and then 7 kilometers due east. What is the magnitude of her resultant displacement? What total distance has she traveled?

2.5 A:

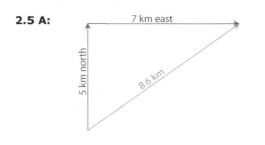

$$a^2 + b^2 = c^2 \rightarrow c = \sqrt{a^2 + b^2} = \sqrt{(5km)^2 + (7km)^2} = 8.6km$$

The hiker's resultant displacement is 8.6 km north of east

The hiker's distance traveled is 12 kilometers.

Notice how for the exact same motion, distance and displacement have significantly different values based on their scalar or vector nature. Understanding the similarities (and differences) between these concepts is an important step toward understanding kinematics.

Speed and Velocity

Knowing only an object's distance and displacement doesn't tell the whole story. Going back to the deer example, there's a significant difference in the picture of the deer's afternoon if the deer's travels occurred over 5 minutes (300 seconds) as opposed to over 50 minutes (3000 seconds).

How exactly does the picture change? In order to answer that question, you'll need to understand some new concepts – average speed and average velocity.

Average speed, given the symbol $\bar{v}$, is defined as distance traveled divided by time, and it tells you the rate at which an object's distance traveled changes. When applying the formula, you must make sure that x is used to represent distance traveled.

$$\bar{v} = \frac{x}{t}$$

2.6 Q: A deer walks 1300 m east to a creek for a drink. The deer then walked 500 m west to the berry patch for dinner, before running 300 m west when startled by a loud raccoon. What is the deer's average speed if the entire trip took 600 seconds (10 minutes)?

2.6 A: $\bar{v} = \dfrac{x}{t} = \dfrac{2100m}{600s} = 3.5\,^{m}\!/_{s}$

Average velocity, also given the symbol $\bar{v}$, is defined as displacement, or change in position, over time. It tells you the rate at which an object's displacement, or position, changes. To calculate the vector quantity average velocity, you divide the vector quantity displacement by time. Note that if you want to find instantaneous speed or velocity, you must take the limit as the time interval gets extremely small (approaches 0).

$$\bar{v} = \frac{x - x_0}{t} = \frac{\Delta x}{t}$$

2.7 Q: A deer walks 1300 m east to a creek for a drink. The deer then walked 500 m west to the berry patch for dinner, before running 300 m west when startled by a loud raccoon. What is the deer's average velocity if the entire trip took 600 seconds (10 minutes)?

2.7 A: $\bar{v} = \dfrac{\Delta x}{t} = \dfrac{500m}{600s} = 0.83\,^{m}\!/_{s}$ east

---

Again, notice how you get very different answers for average speed compared to average velocity. The difference is realizing that distance and speed are scalars, and displacement and velocity are vectors. One way to help you remember these: **s**peed is a **s**calar, and **v**elocity is a **v**ector.

2.8 Q: Chuck the hungry squirrel travels 4m east and 3m north in search of an acorn. The entire trip takes him 20 seconds. Find: Chuck's distance traveled, Chuck's displacement, Chuck's average speed, and Chuck's average velocity.

2.8 A:

$$\Delta x \text{ (distance)} = 4m + 3m = 7m$$

$$\Delta x \text{ (displacement)} = \sqrt{(4m)^2 + (3m)^2}$$

$$\Delta x \text{ (displacement)} = 5m \text{ northeast}$$

$$\bar{v} \text{ (avg. speed)} = \frac{x}{t} = \frac{7m}{20s} = 0.35 \text{ }^m/_s$$

$$\bar{v} \text{ (avg. velocity)} = \frac{\Delta x}{t} = \frac{5m}{20s} = 0.25 \text{ }^m/_s \text{ northeast}$$

2.9 Q: On a highway, a car is driven 80 kilometers during the first 1.00 hour of travel, 50 kilometers during the next 0.50 hour, and 40 kilometers in the final 0.50 hour. What is the car's average speed for the entire trip?

(1) 45 km/h

(2) 60 km/h

(3) 85 km/h

(4) 170 km/h

2.9 A: (3) $\bar{v} = \frac{x}{t} = \frac{170km}{2h} = 85 \text{ }^{km}/_h$

2.10 Q: A person walks 150 meters due east and then walks 30 meters due west. The entire trip takes the person 10 minutes. Determine the magnitude and the direction of the person's total displacement.

2.10 A: 120m due east

Acceleration

So you're starting to get a pretty good understanding of motion. But what would the world be like if velocity never changed? Objects at rest would remain at rest. Objects in motion would remain in motion at a constant speed and direction. Kinetic energy would never change (recall $KE = \frac{1}{2}mv^2$?) It'd make for a pretty boring world. Thankfully, velocity can change, and this change in velocity leads to an **acceleration**.

More accurately, acceleration is the rate at which velocity changes. You can write this as:

$$a = \frac{\Delta v}{t} = \frac{v - v_0}{t}$$

This indicates that the change in velocity divided by the time interval gives you the acceleration. Much like displacement and velocity, acceleration is a vector – it has a direction. Further, the units of acceleration are meters per second per second, or [m/s²]. Although it sounds complicated, all the units mean is that velocity changes at the rate of one meter per second, every second. So an object starting at rest and accelerating at 2 m/s² would be moving at 2 m/s after one second, 4 m/s after two seconds, 6 m/s after three seconds, and so on.

Of special note is the symbolism for Δv. The delta symbol (Δ) indicates a change in a quantity, which is always the initial quantity subtracted from the final quantity. For example:

$$\Delta v = v - v_0$$

2.11 Q: Monty the Monkey accelerates uniformly from rest to a velocity of 9 m/s in a time span of 3 seconds. Calculate Monty's acceleration.

2.11 A: $a = \dfrac{\Delta v}{t} = \dfrac{v - v_0}{t} = \dfrac{9\,{}^m\!/\!_s - 0\,{}^m\!/\!_s}{3s} = 3\,{}^m\!/\!_{s^2}$

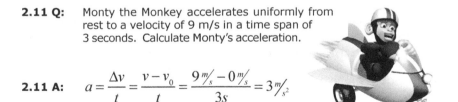

2.12 Q: Velocity is to speed as displacement is to
(1) acceleration
(2) time
(3) momentum
(4) distance

2.12 A: (4) distance. Velocity is the vector equivalent of speed, and displacement is the vector equivalent of distance.

The definition of acceleration can be rearranged to provide a relationship between velocity, acceleration and time as follows:

$$a = \frac{\Delta v}{t} = \frac{v - v_0}{t}$$

$$v - v_0 = at$$

$$v = v_0 + at$$

2.13 Q: The instant before a batter hits a 0.14-kilogram baseball, the velocity of the ball is 45 meters per second west. The instant after the batter hits the ball, the ball's velocity is 35 meters per second east. The bat and ball are in contact for 1.0×10^{-2} second. Determine the magnitude and direction of the average acceleration of the baseball while it is in contact with the bat.

2.13 A: Given: Find:

$v_0 = -45 \,{}^m\!/_s$ a

$v = 35 \,{}^m\!/_s$

$t = 1 \times 10^{-2} s$

$a = \frac{\Delta v}{t} = \frac{v - v_0}{t} = \frac{35\,{}^m\!/_s - (-45\,{}^m\!/_s)}{1 \times 10^{-2} s}$

$a = 8000 \,{}^m\!/_{s^2}$ east

Because acceleration is a vector and has direction, it's important to realize that positive and negative values for acceleration indicate direction only. Take a look at some examples. First, an acceleration of zero implies an object moves at a constant velocity, so a car traveling at 30 m/s east with zero acceleration remains in motion at 30 m/s east.

If the car starts at rest and the car is given a positive acceleration of 5 m/s² east, the car speeds up as it moves east, going faster and faster each second. After one second, the car is traveling 5 m/s. After two seconds, the car travels 10 m/s. After three seconds, the car travels 15 m/s, and so on.

But what happens if the car starts with a velocity of 15 m/s east, and it accelerates at a rate of -5 m/s² (or equivalently, 5 m/s² west)? The car will slow down as it moves to the east until its velocity becomes zero, then it will speed up as it continues to the west.

Positive accelerations don't necessarily indicate an object speeding up, and negative accelerations don't necessarily indicate an object slowing down. In one dimension, for example, if you call east the positive direction, a negative acceleration would indicate an acceleration vector pointing to the west. If the object is moving to the east (has a positive velocity), the negative acceleration would indicate the object is slowing down. If, however, the object is moving to the west (has a negative velocity), the negative acceleration would indicate the object is speeding up as it moves west.

Exasperating, isn't it? Putting it much more simply, if acceleration and velocity have the same sign (vectors in the same direction), the object is speeding up. If acceleration and velocity have opposite signs (vectors in opposite directions), the object is slowing down.

Particle Diagrams

Graphs and diagrams are terrific tools for understanding physics, and they are especially helpful for studying motion, a phenomenon that we are used to perceiving visually. Particle diagrams, sometimes referred to as ticker-tape diagrams or dot diagrams, show the position or displacement of an object at evenly spaced time intervals.

Think of a particle diagram like an oil drip pattern... if a car has a steady oil drip, where one drop of oil falls to the ground every second, the pattern of the oil droplets on the ground could represent the motion of the car with respect to time. By examining the oil drop pattern, a bystander could draw conclusions about the displacement, velocity, and acceleration of the car, even if he wasn't able to watch the car drive by! The oil drop pattern is known as a particle, or ticker-tape, diagram.

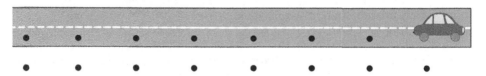

From the particle diagram above you can see that the car was moving either to the right or the left, and since the drops are evenly spaced, you can say with certainty that the car was moving at a constant velocity, and since velocity isn't changing, acceleration must be 0. So what would the particle diagram look like if the car was accelerating to the right? Take a look below and see!

The oil drops start close together on the left, and get further and further apart as the object moves toward the right. Of course, this pattern could also have been produced by a car moving from right to left, beginning with a high velocity at the right and slowing down as it moves toward the left. Because the velocity vector (pointing to the left) and the acceleration vector (pointing to the right) are in opposite directions, the object slows down.

This is a case where, if you called to the right the positive direction, the car would have a negative velocity, a positive acceleration, and it would be slowing down. Check out the resulting particle diagram below!

Can you think of a case in which the car could have a negative velocity and a negative acceleration, yet be speeding up? Draw a picture of the situation!

Position-Time (x-t) Graphs

As you've observed, particle diagrams can help you understand an object's motion, but they don't always tell you the whole story. You'll have to investigate some other types of motion graphs to get a clearer picture.

The position-time graph shows the displacement (or, in the case of scalar quantities, distance) of an object as a function of time. Positive displacements indicate the object's position is in the positive direction from its starting point, while negative displacements indicate the object's position is opposite the positive direction. Let's look at a few examples.

Suppose Cricket the Wonder Dog wanders away from her house at a constant velocity of 1 m/s, stopping only when she's 5m away (which, of course, takes 5 seconds). She then decides to take a short five second rest in the grass. After her five second rest, she hears the dinner bell ring, so she runs back to the house at a speed of 2 m/s. The position-time graph for her motion would look something like this:

As you can see from the plot, Cricket's displacement begins at zero meters at time zero. Then, as time progresses, Cricket's position changes at a rate of 1 m/s, so that after one second, Cricket is one meter away from her starting point. After two seconds, she's two meters away, and so forth, until she reaches her maximum displacement of five meters from her starting point at a time of five seconds (her position is now x=5m). Cricket then remains at that position for 5 seconds while she takes a rest. Following her rest, at time t=10 seconds, Cricket hears the dinner bell and races back to the house at a speed of 2 m/s, so the graph ends when Cricket returns to her starting

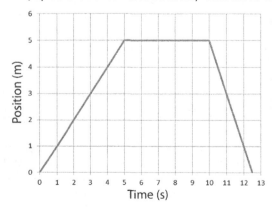

point at the house, a total distance traveled of 10m, and a total displacement of zero meters.

As you look at the position-time graph, notice that at the beginning, when
Cricket is moving in a positive direction, the graph has a positive slope. When
the graph is flat (has a zero slope), Cricket is not moving. When the graph
has a negative slope, Cricket is moving in the negative direction. It's also easy
to see that the steeper the slope of the graph, the faster Cricket is moving.

2.14 Q: The graph below represents the displacement of an object moving in a straight line as a function of time.

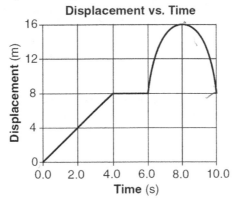

Displacement vs. Time

What was the total distance traveled by the object during the
10-second time interval?

(1) 0 m
(2) 8 m
(3) 16 m
(4) 24 m

2.14 A: (4) Total distance traveled is 8 meters forward from 0 to 4 seconds, then 8 meters forward from 6 to 8 seconds, then 8 meters
backward from 8 to 10 seconds, for a total of 24 meters.

Velocity-Time (v-t) Graphs

Just as important to understanding motion is the velocity-time graph, which
shows the velocity of an object on the y-axis, and time on the x-axis. Positive values indicate velocities in the positive direction, while negative values indicate velocities in the opposite direction. In reading these graphs,
it's important to realize that a straight horizontal line indicates the object
maintaining a constant velocity – it can still be moving, its velocity just isn't
changing. A value of 0 on the v-t graph indicates the object has come to a
stop. If the graph crosses the x-axis, the object was moving in one direction, came to a stop, and switched the direction of its motion. Let's look at
the v-t graph for Cricket the Wonderdog's Adventure:

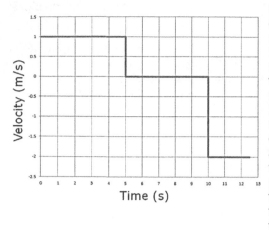

Time (s)

For the first five seconds of Cricket's journey, you can see she maintains a constant velocity of 1 m/s. Then, when she stops to rest, her velocity changes to zero for the duration of her rest. Finally, when she races back to the house for dinner, she maintains a negative velocity of 2 m/s. Because velocity is a vector, the negative sign indicates that Cricket's velocity is in the opposite direction (initially the direction away from the house was positive, so back toward the house must be negative!)

As I'm sure you can imagine, the position-time graph of an object's motion and the v-t graph of an object's motion are closely related. You'll explore these relationships next.

Graph Transformations

In looking at a position-time graph, the faster an object's position/displacement changes, the steeper the slope of the line. Since velocity is the rate at which an object's position changes, the slope of the position-time graph at any given point in time gives you the velocity at that point in time. You can obtain the slope of the position-time graph using the following formula:

$$slope = \frac{rise}{run} = \frac{y_2 - y_1}{x_2 - x_1}$$

Realizing that the rise in the graph is actually Δx, and the run is Δt, you can substitute these variables into the slope equation to find:

$$slope = \frac{rise}{run} = \frac{\Delta x}{\Delta t} = v$$

With a little bit of interpretation, it's easy to show that the slope is really just change in position over time, which is the definition of velocity. Put directly, the slope of the position-time graph gives you the velocity.

Of course, it only makes sense that if you can determine velocity from the position-time graph, you should be able to work backward to determine change in position (displacement) from the v-t graph. If you have a v-t graph, and you want to know how much an object's position changed in a time interval, take the area under the curve within that time interval.

Chapter 2: Kinematics

So, if taking the slope of the position-time graph gives you the rate of change of position, which is called velocity, what do you get when you take the slope of the v-t graph? You get the rate of change of velocity, which is called acceleration! The slope of the v-t graph, therefore, tells you the acceleration of an object.

$$slope = \frac{rise}{run} = \frac{\Delta v}{\Delta t} = a$$

2.15 Q: The graph below represents the motion of a car during a 6.0-second time interval.

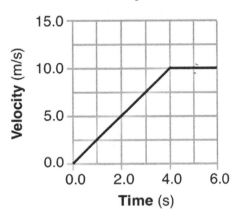

Velocity vs. Time

(A) What is the total distance traveled by the car during this 6-second interval?

(B) What is the acceleration of the car at t = 5 seconds?

2.15 A: (A) distance = area under graph

distance = $Area_{triangle}$ + $Area_{rectangle}$

distance = $\frac{1}{2}bh + lw$

distance = $\frac{1}{2}(4s)(10\,{}^m\!/\!_s) + (2s)(10\,{}^m\!/\!_s)$

distance = 40m

(B) acceleration = slope at t=5 seconds = 0 because graph is flat at t=5 seconds.

2.16 Q: The graph below represents the velocity of an object traveling in a straight line as a function of time.

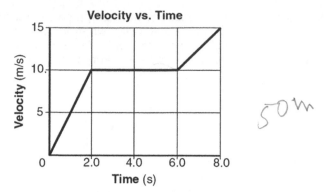

Velocity vs. Time

Determine the magnitude of the total displacement of the object at the end of the first 6.0 seconds.

2.16 A: displacement = area under graph

displacement = $Area_{triangle} + Area_{rectangle}$

displacement = $\frac{1}{2}bh + lw$

displacement = $\frac{1}{2}(2s)(10\,\%_s) + (4s)(10\,\%_s)$

displacement = $50m$

2.17 Q: The graph below shows the velocity of a race car moving along a straight line as a function of time.

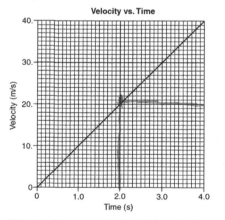

What is the magnitude of the displacement of the car from t = 2.0 seconds to t = 4.0 seconds?

(1) 20 m

(2) 40 m

(3) 60 m

(4) 80 m

2.17 A: (3) 60 m

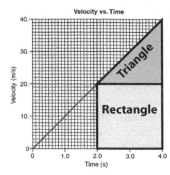

displacement = area under graph

displacement = $Area_{triangle} + Area_{rectangle}$

displacement = $\frac{1}{2}bh + lw$

displacement = $\frac{1}{2}(2s)(20\,{}^{m}\!/_{s}) + (2s)(20\,{}^{m}\!/_{s})$

displacement = $60m$

2.18 Q: The displacement-time graph below represents the motion of a cart initially moving forward along a straight line.

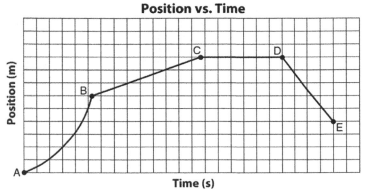

During which interval is the cart moving forward at constant speed?

(1) AB
(2) BC
(3) CD
(4) DE

2.18 A: (2) The slope of the position-time graph is constant and positive during interval BC, therefore the velocity of the cart must be constant and positive in that interval.

Acceleration-Time (a-t) Graphs

Much like velocity, you can make a graph of acceleration vs. time by plotting the rate of change of an object's velocity (its acceleration) on the y-axis, and placing time on the x-axis.

When you took the slope of the position-time graph, you obtained the object's velocity. In the same way, taking the slope of the v-t graph gives you the object's acceleration. Going the other direction, when you analyzed the v-t graph, you found that taking the area under the v-t graph provided you with information about the object's change in position. In similar fashion, taking the area under the a-t graph tells you how much an object's velocity changes.

Putting it all together, you can go from position-time to velocity-time by taking the slope, and you can go from velocity-time to acceleration-time by taking the slope. Or, going the other direction, the area under the acceleration-time curve gives you an object's change in velocity, and the area under the velocity-time curve gives you an object's change in position.

Graphical Analysis of Motion
How do I move from one type of graph to another?

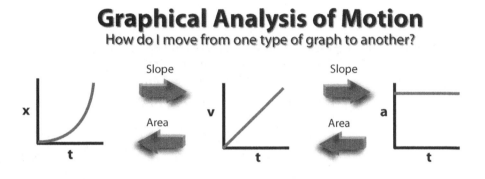

2.19 Q: Which graph best represents the motion of a block accelerating uniformly down an inclined plane?

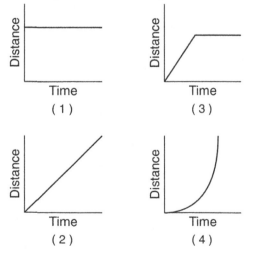

2.19 A: (4) If the block accelerates uniformly, it must have a constant acceleration. This means the v-t graph must be a straight line, since its slope, which is equal to its acceleration, must be constant. Therefore, the v-t graph must look something like the graph at right. The slope of the position-time graph, which gives velocity, must be constantly increasing to give the v-t graph at right. The only answer choice with a constantly increasing slope is (4), so (4) must be the answer!

2.20 Q: A student throws a baseball vertically upward and then catches it. If vertically upward is considered to be the positive direction, which graph best represents the relationship between velocity and time for the baseball?

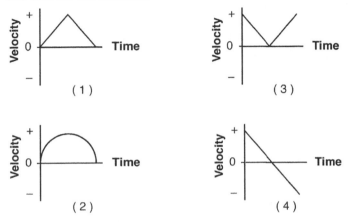

2.20 A: (4) If up is considered the positive direction, and the baseball is thrown upward to start its motion, it begins with a positive velocity. At its highest point, its vertical velocity becomes 0. Then, it speeds up as it comes down, so it obtains a larger and larger negative velocity.

2.21 Q: A cart travels with a constant nonzero acceleration along a straight line. Which graph best represents the relationship between the distance the cart travels and time of travel?

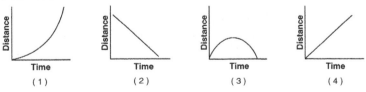

2.21 A: (1) A constant acceleration is caused by a linearly increasing velocity. Since velocity is obtained from the slope of the position-time graph, the position-time graph must be continually increasing to provide the correct v-t graph.

Kinematic Equations

Motion graphs such as the position-time, velocity-time, and acceleration-time graphs are terrific tools for understanding motion. However, there are times when graphing motion may not be the most efficient or effective way of understanding the motion of an object. To assist in these situations, you can add a set of problem-solving equations to your physics toolbox, known as the **kinematic equations**. These equations can help you solve for key variables describing the motion of an object when you have a constant acceleration. Once you know the value of any three variables, you can use the kinematic equations to solve for the other two!

Variable	Meaning
v_0	initial velocity
v	final velocity
Δx	displacement
a	acceleration
t	time

$$v = v_0 + at$$

$$v^2 = v_0^2 + 2a\Delta x$$

$$\Delta x = v_0 t + \tfrac{1}{2}at^2$$

In using these equations to solve motion problems, it's important to take care in setting up your analysis before diving in to a solution. Key steps to solving kinematics problems include:

1. Labeling your analysis for horizontal (x-axis) or vertical (y-axis) motion.
2. Choosing and indicating a positive direction (typically the direction of initial motion).
3. Creating a motion analysis table (v_0, v, Δx, a, t). Note that Δx is a change in position, or displacement, and can be re-written as $x-x_0$.
4. Using what you know about the problem to fill in your "givens" in the table.
5. Once you know three items in the table, use kinematic equations to solve for any unknowns.
6. Verify that your solution makes sense.

Take a look at a sample problem to see how this strategy can be employed.

2.22 Q: A race car starting from rest accelerates uniformly at a rate of 4.90 meters per second². What is the car's speed after it has traveled 200 meters?

(1) 1960 m/s

(2) 62.6 m/s

(3) 44.3 m/s

(4) 31.3 m/s

2.22 A: Step 1: Horizontal Problem

Step 2: Positive direction is direction car starts moving.

Step 3 & 4:

Variable	Value
v_0	0 m/s
v	FIND
Δx	200 m
a	4.90 m/s²
t	?

Step 5: Choose a kinematic equation that includes the given information and the information sought, and solve for the unknown showing the initial formula, substitution with units, and answer with units.

$$v^2 = v_0^2 + 2a\Delta x$$
$$v^2 = (0\,{}^m\!/_s)^2 + 2(4.90\,{}^m\!/_{s^2})(200m)$$
$$v^2 = 1960\,{}^{m^2}\!/_{s^2}$$
$$v^2 = \sqrt{1960\,{}^{m^2}\!/_{s^2}} = 44.3\,{}^m\!/_s$$

Step 6: (3) 44.3 m/s is one of the given answer choices, and is a reasonable speed for a race car (44.3 m/s is approximately 99 miles per hour).

This strategy also works for vertical motion problems.

2.23 Q: An astronaut standing on a platform on the Moon drops a hammer. If the hammer falls 6.0 meters vertically in 2.7 seconds, what is its acceleration?

(1) 1.6 m/s²
(2) 2.2 m/s²
(3) 4.4 m/s²
(4) 9.8 m/s²

2.23 A: Step 1: Vertical Problem

Step 2: Positive direction is down (direction hammer starts moving.)

Step 3 & 4: Note that a dropped object has an initial vertical velocity of 0 m/s.

Variable	Value
v_0	0 m/s
v	?
Δy	6 m
a	FIND
t	2.7 s

Step 5: Choose a kinematic equation that includes the given information and the information sought, and solve for the unknown showing the initial formula, substitution with units, and answer with units.

$$\Delta y = v_0 t + \tfrac{1}{2}at^2 \xrightarrow{v_0=0}$$

$$\Delta y = \tfrac{1}{2}at^2$$

$$a = \frac{2\Delta y}{t^2}$$

$$a = \frac{2(6m)}{(2.7s)^2} = 1.6\,m/_{s^2}$$

Step 6: (1) 1.6 m/s² is one of the given answer choices, and is less than the acceleration due to gravity on the surface of the Earth (9.8 m/s²). This answer can be verified further by searching on the Internet to confirm that the acceleration due to gravity on the surface of the moon is indeed 1.6 m/s².

In some cases, you may not be able to solve directly for the "find" quantity. In these cases, you can solve for the other unknown variable first, then choose an equation to give you your final answer.

2.24 Q: A car traveling on a straight road at 15.0 meters per second accelerates uniformly to a speed of 21.0 meters per second in 12.0 seconds. The total distance traveled by the car in this 12.0-second time interval is

(1) 36.0 m

(2) 180 m

(3) 216 m

(4) 252 m

2.24 A: Horizontal Problem, positive direction is forward.

Variable	Value
v_0	15 m/s
v	21 m/s
Δx	FIND
a	?
t	12 s

Can't find Δx directly, find a first.

$$v = v_0 + at$$

$$a = \frac{v - v_0}{t}$$

$$a = \frac{21\,m/_s - 15\,m/_s}{12s} = 0.5\,m/_{s^2}$$

Now solve for Δx.

$$\Delta x = v_0 t + \tfrac{1}{2}at^2$$

$$\Delta x = (15\,{}^m\!/_s)(12s) + \tfrac{1}{2}(0.5\,{}^m\!/_{s^2})(12s)^2$$

$$\Delta x = 216m$$

Check: (3) 216m is a given answer, and is reasonable, as this is greater than the 180m the car would have traveled if remaining at a constant speed of 15 m/s for the 12 second time interval.

2.25 Q: A car initially traveling at a speed of 16 meters per second accelerates uniformly to a speed of 20 meters per second over a distance of 36 meters. What is the magnitude of the car's acceleration?

(1) 0.11 m/s²

(2) 2.0 m/s²

(3) 0.22 m/s²

(4) 9.0 m/s²

2.25 A: (2) $v^2 = v_0^2 + 2a\Delta x$

$$a = \frac{v^2 - v_0^2}{2\Delta x}$$

$$a = \frac{(20\,{}^m\!/_s)^2 - (16\,{}^m\!/_s)^2}{2(36m)} = 2\,{}^m\!/_{s^2}$$

2.26 Q: An astronaut drops a hammer from 2.0 meters above the surface of the Moon. If the acceleration due to gravity on the Moon is 1.62 meters per second², how long will it take for the hammer to fall to the Moon's surface?

(1) 0.62 s

(2) 1.2 s

(3) 1.6 s

(4) 2.5 s

2.26 A: (3) $\Delta y = v_0 t + \tfrac{1}{2}at^2 \xrightarrow{\;v_0 = 0\;}$

$$\Delta y = \tfrac{1}{2}at^2$$

$$t = \sqrt{\frac{2\Delta y}{a}} = \sqrt{\frac{2(2m)}{1.62\,{}^m\!/_{s^2}}} = 1.57s$$

2.27 Q: A car increases its speed from 9.6 meters per second to 11.2 meters per second in 4.0 seconds. The average acceleration of the car during this 4.0-second interval is

(1) 0.40 m/s²

(2) 2.4 m/s²

(3) 2.8 m/s²

(4) 5.2 m/s²

2.27 A: (1) $v = v_0 + at$

$$a = \frac{v - v_0}{t} = \frac{(11.2\,^m/_s) - (9.6\,^m/_s)}{4s} = 0.4\,^m/_{s^2}$$

2.28 Q: A rock falls from rest a vertical distance of 0.72 meters to the surface of a planet in 0.63 seconds. The magnitude of the acceleration due to gravity on the planet is

(1) 1.1 m/s²

(2) 2.3 m/s²

(3) 3.6m/s²

(4) 9.8 m/s²

2.28 A: (3) $\Delta y = v_0 t + \frac{1}{2}at^2 \xrightarrow{\;v_0 = 0\;}$

$$\Delta y = \frac{1}{2}at^2$$

$$a = \frac{2\Delta y}{t^2} = \frac{2(0.72m)}{(0.63s)^2} = 3.6\,^m/_{s^2}$$

2.29 Q: The speed of an object undergoing constant acceleration increases from 8.0 meters per second to 16.0 meters per second in 10 seconds. How far does the object travel during the 10 seconds?

(1) 3.6×10² m

(2) 1.6×10² m

(3) 1.2×10² m

(4) 8.0×10¹ m

2.29 A: (3) Can't find Δx directly, find a first.

$$v = v_0 + at$$

$$a = \frac{v - v_0}{t}$$

$$a = \frac{16\,^m/_s - 8\,^m/_s}{10s} = 0.8\,^m/_{s^2}$$

Now solve for Δx.

$$\Delta x = v_0 t + \tfrac{1}{2} a t^2$$
$$\Delta x = (8 \, ^m\!/_s)(10s) + \tfrac{1}{2}(0.8 \, ^m\!/_{s^2})(10s)^2$$
$$\Delta x = 120m$$

Free Fall

Examination of free-falling bodies dates back to the days of Aristotle. At that time Aristotle believed that more massive objects would fall faster than less massive objects. He believed this in large part due to the fact that when examining a rock and a feather falling from the same height it is clear that the rock hits the ground first. Upon further examination it is clear that Aristotle was incorrect in his hypothesis.

As proof, take a basketball and a piece of paper. Drop them simultaneously from the same height... do they land at the same time? Probably not. Now take that piece of paper and crumple it up into a tight ball and repeat the experiment. Now what do you see happen? You should see that both the ball and the paper land at the same time. Therefore you can conclude that Aristotle's predictions did not account for the effect of air resistance. For the purposes of this course, drag forces such as air resistance will be neglected.

In the 17th century, Galileo Galilei began a re-examination of the motion of falling bodies. Galileo, recognizing that air resistance affects the motion of a falling body, executed his famous thought experiment in which he continuously asked what would happen if the effect of air resistance was removed. Commander David Scott of Apollo 15 performed this experiment while on the moon. He simultaneously dropped a hammer and a feather, and observed that they reached the ground at the same time.

Since Galileo's experiments, scientists have come to a better understanding of how the gravitational pull of the Earth accelerates free-falling bodies. Through experimentation it has been determined that the local **gravitational field strength (g)** on the surface of the Earth is 9.8 N/kg, which further indicates that all objects in free fall (neglecting air resistance) experience an equivalent acceleration of 9.8 m/s^2 toward the center of the Earth. (NOTE: If you move off the surface of the Earth the local gravitational field strength, and therefore the acceleration due to gravity, changes.)

You can look at free-falling bodies as objects being dropped from some height or thrown vertically upward. In this examination you will analyze the motion of each condition.

Objects Falling From Rest

Objects starting from rest have an initial velocity of zero, giving you your first kinematic quantity needed for problem solving. Beyond that, if you call the direction of initial motion (down) positive, the object will have a positive acceleration and speed up as it falls.

An important first step in analyzing objects in free fall is decid-ing which direction along the y-axis you are going to call posi-tive and which direction will therefore be negative. Although you can set your positive direction any way you want and get the correct answer, following the hints below can simplify your work to reach the correct answer consistently.

1. Identify the direction of the object's initial motion and assign that as the positive direction. In the case of a dropped object, the positive y-direction will point toward the bottom of the paper.
2. With the axis identified you can now identify and write down your given kinematic information. Don't forget that a dropped object has an initial velocity of zero.
3. Notice the direction the vector arrows are drawn — if the velocity and acceleration point in the same direction, the object speeds up. If they point in opposite directions, the object slows down.

2.30 Q: What is the speed of a 2.5-kilogram mass after it has fallen freely from rest through a distance of 12 meters?

(1) 4.8 m/s

(2) 15 m/s

(3) 30 m/s

(4) 43 m/s

2.30 A: Vertical Problem: Declare down as the positive direction. This means that the acceleration, which is also down, is a positive quantity.

Variable	Value
v_0	0 m/s
v	FIND
Δy	12 m
a	9.8 m/s²
t	?

$$v^2 = v_0^2 + 2a\Delta y$$
$$v^2 = (0\,\text{m}/\text{s})^2 + 2(9.8\,\text{m}/\text{s}^2)(12m)$$
$$v^2 = 235\,\text{m}^2/\text{s}^2$$
$$v = \sqrt{235\,\text{m}^2/\text{s}^2} = 15.3\,\text{m}/\text{s}$$

Correct answer is (2) 15 m/s.

2.31 Q: How far will a brick starting from rest fall freely in 3.0 seconds?
(1) 15 m
(2) 29 m
(3) 44 m
(4) 88 m

2.31 A: (3) 44m

Variable	Value
v_0	0 m/s
v	?
Δy	FIND
a	9.8 m/s²
t	3 s

$$\Delta y = v_0 t + \tfrac{1}{2}at^2$$
$$\Delta y = (0\,\text{m}/\text{s})(3s) + \tfrac{1}{2}(9.8\,\text{m}/\text{s}^2)(3s)^2$$
$$\Delta y = 44m$$

2.32 Q: A ball dropped from rest falls freely until it hits the ground with a speed of 20 meters per second. The time during which the ball is in free fall is approximately
(1) 1 s
(2) 2 s
(3) 0.5 s
(4) 10 s

2.32 A:

Variable	Value
v_0	0 m/s
v	20 m/s
Δy	?
a	9.8 m/s²
t	FIND

$$v = v_0 + at$$

$$t = \frac{v - v_0}{a}$$

$$t = \frac{20\,^m/_s - 0\,^m/_s}{9.8\,^m/_{s^2}} = 2.04s$$

(2) 2 s

Objects Launched Upward

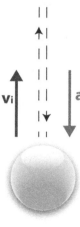

Examining the motion of an object launched vertically upward is done in much the same way you examined the motion of an object falling from rest. The major difference is that you have to look at two segments of its motion instead of one: both up *and* down.

Before you get into establishing a frame of reference and working through the quantitative analysis, you must build a solid conceptual understanding of what is happening while the ball is in the air. Consider the ball being thrown vertically into the air as shown in the diagram.

In order for the ball to move upwards its initial velocity must be greater than zero. As the ball rises, its velocity decreases until it reaches its maximum height, where it stops, and then begins to fall. As the ball falls, its speed increases. In other words, the ball is accelerating the entire time it is in the air, both on the way up, at the instant it stops at its highest point, and on the way down.

The cause of the ball's acceleration is gravity. The entire time the ball is in the air, its acceleration is 9.8 m/s^2 down provided this occurs on the surface of the Earth. Note that the acceleration can be either 9.8 m/s^2 or -9.8 m/s^2. The sign of the acceleration depends on the direction you declared as positive, but in all cases the direction of the acceleration due to gravity is down, toward the center of the Earth.

You have already established the ball's acceleration for the entire time it is in the air is 9.8 m/s^2 down. This acceleration causes the ball's velocity to decrease at a constant rate until it reaches maximum altitude, at which point it turns around and starts to fall. In order to turn around, the ball's velocity must pass through zero. Therefore, at maximum altitude the velocity of the ball must be zero.

2.33 Q: A ball thrown vertically upward reaches a maximum height of 30 meters above the surface of Earth. At its maximum height, the speed of the ball is

(1) 0 m/s
(2) 3.1 m/s
(3) 9.8 m/s
(4) 24 m/s

2.33 A: (1) 0 m/s. The instantaneous speed of any projectile at its maximum height is zero.

Because gravity provides the same acceleration to the ball on the way up (slowing it down) as on the way down (speeding it up), the time to reach maximum altitude is the same as the time to return to its launch position. In similar fashion, the initial velocity of the ball on the way up will equal the velocity of the ball at the instant it reaches the point from which it was launched on the way down. Put another way, the time to go up is equal to the time to go down, and the initial velocity up is equal to the final velocity down (assuming the object begins and ends at the same height above ground).

Now that a conceptual understanding of the ball's motion has been established, you can work toward a quantitative solution. Following the rule of thumb established previously, you can start by assigning the direction the ball begins to move as positive. Remember that assigning positive and negative directions are completely arbitrary. You have the freedom to assign them how you see fit. Once you assign them, however, don't change them.

Once this positive reference direction has been established, all other velocities and displacements are assigned accordingly. For example, if up is the positive direction, the acceleration due to gravity will be negative, because the acceleration due to gravity points down, toward the center of the Earth. At its highest point, the ball will have a positive displacement, and will have a zero displacement when it returns to its starting point. If the ball isn't caught, but continues toward the Earth past its starting point, it will have a negative displacement.

A "trick of the trade" to solving free fall problems involves symmetry. The time an object takes to reach its highest point is equal to the time it takes to return to the same vertical position. The speed with which the projectile begins its journey upward is equal to the speed of the projectile when it returns to the same height (although, of course, its velocity is in the opposite direction). If you want to simplify the problem, vertically, at its highest point, the vertical velocity is 0. This added information can assist you in filling out your vertical motion table. If you cut the object's motion in half, you can simplify your problem solving – but don't forget that if you want the total time in the air, you must double the time it takes for the object to rise to its highest point.

2.34 Q: A basketball player jumped straight up to grab a rebound. If she was in the air for 0.80 seconds, how high did she jump?

(1) 0.50 m

(2) 0.78 m

(3) 1.2 m

(4) 3.1 m

2.34 A: Define up as the positive y-direction. Note that if basketball player is in the air for 0.80 seconds, she reaches her maximum height at a time of 0.40 seconds, at which point her velocity is zero.

Variable	Value
v_0	?
v	0 m/s
Δy	FIND
a	-9.8 m/s²
t	0.40 s

Can't solve for Δx directly with given information, so find v_0 first.

$$v = v_0 + at$$

$$v_0 = v - at$$

$$v_0 = 0 - (-9.8\,\tfrac{m}{s^2})(0.40s) = 3.92\,\tfrac{m}{s}$$

Now with v_0 known, solve for Δx.

$$\Delta y = v_0 t + \tfrac{1}{2}at^2$$

$$\Delta y = (3.92\,\tfrac{m}{s})(0.40s) + \tfrac{1}{2}(-9.8\,\tfrac{m}{s^2})(0.40s)^2$$

$$\Delta y = 0.78m$$

Correct answer is (2) 0.78 m. This is a reasonable height for a basketball player to jump.

2.35 Q: Which graph best represents the relationship between the acceleration of an object falling freely near the surface of Earth and the time that it falls?

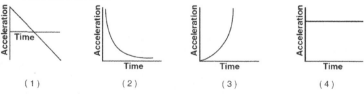

2.35 A: (4) The acceleration due to gravity is a constant 9.8 m/s² down on the surface of the Earth.

2.36 Q: A ball is thrown straight downward with a speed of 0.50 meter per second from a height of 4.0 meters. What is the speed of the ball 0.70 seconds after it is released?

(1) 0.50 m/s

(2) 7.4 m/s

(3) 9.8 m/s

(4) 15 m/s

2.36 A: (2) 7.4 m/s. Note that in filling out the kinematics table, the height of 4 meters is not the displacement of the ball, but is extra unneeded information.

Variable	Value
v_0	0.50 m/s
v	FIND
Δy	?
a	9.8 m/s²
t	0.70 s

$$v = v_0 + at$$
$$v = 0.50 \, ^m\!/_s + (9.8 \, ^m\!/_{s^2})(0.70s)$$
$$v = 7.4 \, ^m\!/_s$$

2.37 Q: A baseball dropped from the roof of a tall building takes 3.1 seconds to hit the ground. How tall is the building? [Neglect friction.]

(1) 15 m

(2) 30 m

(3) 47 m

(4) 94 m

2.37 A: (3) 47 m

$$\Delta y = v_0 t + \tfrac{1}{2}at^2$$
$$\Delta y = (0 \, ^m\!/_s)(3.1s) + \tfrac{1}{2}(9.8 \, ^m\!/_{s^2})(3.1s)^2$$
$$\Delta y = 47m$$

2.38 Q: A 0.25-kilogram baseball is thrown upward with a speed of 30 meters per second. Neglecting friction, the maximum height reached by the baseball is approximately

(1) 15 m

(2) 46 m

(3) 74 m

(4) 92 m

2.38 A: (2) 46m

$$v^2 = v_0^2 + 2a\Delta y$$

$$\Delta y = \frac{v^2 - v_0^2}{2a}$$

$$\Delta y = \frac{(0\,^m/_s)^2 - (30\,^m/_s)^2}{2(-9.8\,^m/_{s^2})} = 45.9m$$

Projectile Motion

Projectile motion problems, or problems of an object launched in both the x- and y- directions, can be analyzed using the physics you already know. The key to solving these types of problems is realizing that the horizontal component of the object's motion is independent of the vertical component of the object's motion. Since you already know how to solve horizontal and vertical kinematics problems, all you have to do is put the two results together!

Start these problems by making separate motion tables for vertical and horizontal motion. Vertically, the setup is the same for projectile motion as it is for an object in free fall. Horizontally, gravity only pulls an object down, it never pulls or pushes an object horizontally, therefore the horizontal acceleration of any projectile is zero. If the acceleration horizontally is zero, velocity must be constant, therefore v_0 horizontally must equal v horizontally. Finally, to tie the problem together, realize that the time the projectile is in the air vertically must be equal to the time the projectile is in the air horizontally.

When an object is launched or thrown completely horizontally, such as a rock thrown horizontally off a cliff, the initial velocity of the object is its initial horizontal velocity. Because horizontal velocity doesn't change, this velocity is also the object's final horizontal velocity, as well as its average horizontal velocity. Further, the initial vertical velocity of the projectile is zero. This means that you could hurl an object 1000 m/s horizontally off a cliff, and simultaneously drop an object off the cliff from the same height, and they will both reach the ground at the same time (even though the hurled object has traveled a greater distance).

2.39 Q: Fred throws a baseball 42 m/s horizontally from a height of two meters. How far will the ball travel before it reaches the ground?

2.39 A: To solve this problem, you must first find how long the ball will remain in the air. This is a vertical motion problem.

VERTICAL MOTION TABLE

Variable	Value
v_0	0 m/s
v	?
Δy	2 m
a	9.8 m/s²
t	FIND

$$\Delta y = v_0 t + \tfrac{1}{2} at^2$$

$$\Delta y = \tfrac{1}{2} at^2$$

$$t = \sqrt{\frac{2\Delta y}{a}}$$

$$t = \sqrt{\frac{2(2m)}{9.8 \, m/_{s^2}}} = 0.639s$$

Now that you know the ball is in the air for 0.639 seconds, you can find how far it travels horizontally before reaching the ground. This is a horizontal motion problem, in which the acceleration is 0 (nothing is causing the ball to accelerate horizontally.) Because the ball doesn't accelerate, its initial velocity is also its final velocity, which is equal to its average velocity.

HORIZONTAL MOTION TABLE

Variable	Value
v_0	42 m/s
v	42 m/s
Δx	FIND
a	0 m/s²
t	0.639 s

$$\bar{v} = \frac{\Delta x}{t}$$

$$\Delta x = \bar{v} t = (42 \, m/_s)(0.639s) = 26.8m$$

You can therefore conclude that the baseball travels 26.8 meters horizontally before reaching the ground.

2.40 Q: The diagram below represents the path of a stunt car that is driven off a cliff, neglecting friction.

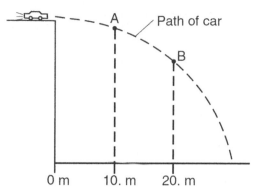

Path of car

0 m   10. m   20. m

Compared to the horizontal component of the car's velocity at point A, the horizontal component of the car's velocity at B is

(1) smaller

(2) greater

(3) the same

2.40 A: (3) the same. The car's horizontal acceleration is zero, therefore the horizontal velocity remains constant.

2.41 Q: A 0.2-kilogram red ball is thrown horizontally at a speed of 4 meters per second from a height of 3 meters. A 0.4-kilogram green ball is thrown horizontally from the same height at a speed of 8 meters per second. Compared to the time it takes the red ball to reach the ground, the time it takes the green ball to reach the ground is

(1) one-half as great

(2) twice as great

(3) the same

(4) four times as great

2.41 A: (3) the same. Both objects are thrown horizontally from the same height. Because horizontal motion and vertical motion are independent, both objects have the same vertical motion (they both start with an initial vertical velocity of 0, have the same acceleration of 9.8 m/s² down, and both travel the same vertical distance). Therefore, the two objects reach the ground in the same amount of time.

2.42 Q: A ball is thrown horizontally at a speed of 24 meters per second from the top of a cliff. If the ball hits the ground 4.0 seconds later, approximately how high is the cliff?

(1) 6.0 m

(2) 39 m

(3) 78 m

(4) 96 m

2.42 A: (3) 78m

$$\Delta y = v_0 t + \tfrac{1}{2} at^2$$
$$\Delta y = \tfrac{1}{2} at^2$$
$$\Delta y = \tfrac{1}{2}(9.8 \tfrac{m}{s^2})(4s)^2$$
$$\Delta y = 78m$$

2.43 Q: Projectile A is launched horizontally at a speed of 20 meters per second from the top of a cliff and strikes a level surface below, 3.0 seconds later. Projectile B is launched horizontally from the same location at a speed of 30 meters per second. The time it takes projectile B to reach the level surface is

(1) 4.5 s

(2) 2.0 s

(3) 3.0 s

(4) 10 s

2.43 A: (3) 3.0 s. They both take the same time to reach the ground because they both travel the same distance vertically, and they both have the same vertical acceleration (9.8 m/s² down) and initial vertical velocity (zero).

Angled Projectiles

For objects launched at an angle, you have to do a little more work to determine the initial velocity in both the horizontal and vertical directions. For example, if a football is kicked with an initial velocity of 40 m/s at an angle of 30° above the horizontal, you need to break the initial velocity vector up into x- and y-components in the same manner as covered in the components of vectors math review section.

Then, use the components for your initial velocities in your horizontal and vertical tables. Finally, don't forget that symmetry of motion also   applies to the parabola of projectile motion. For objects launched and landing at the same height, the launch angle is equal to the landing angle. The launch velocity is equal to the landing velocity. And if you want an object to travel the maximum possible horizontal distance (or range), launch it at an angle of 45°.

2.44 Q: Herman the human cannonball is launched from level ground at an angle of 30° above the horizontal with an initial velocity of 26 m/s. How far does Herman travel horizontally before reuniting with the ground?

2.44 A: The first step in solving this type of problem is to determine Herman's initial horizontal and vertical velocity. You do this by breaking up his initial velocity into vertical and horizontal components:

$$v_{0_x} = v_0 \cos(\theta) = (26\,{}^m\!/_s)\cos(30°) = 22.5\,{}^m\!/_s$$

$$v_{0_y} = v_0 \sin(\theta) = (26\,{}^m\!/_s)\sin(30°) = 13\,{}^m\!/_s$$

Next, analyze Herman's vertical motion to find out how long he is in the air. You can analyze his motion on the way up, find the time, and double that to find his total time in the air:

VERTICAL MOTION TABLE

Variable	Value
v_0	13 m/s
v	0 m/s
Δy	?
a	-9.8 m/s²
t	FIND

$$v = v_0 + at$$

$$t = \frac{v - v_0}{a}$$

$$t_{up} = \frac{(0 - 13\,{}^m\!/_s)}{-9.8\,{}^m\!/_{s^2}} = 1.33s$$

$$t_{total} = 2 \times t_{up} = 2.65s$$

Now that you know Herman was in the air 2.65s, you can find how far he moved horizontally, using his initial horizontal velocity of 22.5 m/s.

HORIZONTAL MOTION TABLE

Variable	Value
v_0	22.5 m/s
v	22.5 m/s
Δx	FIND
a	0
t	2.65 s

$$\bar{v} = \frac{\Delta x}{t}$$

$$\Delta x = \bar{v}t$$

$$\Delta x = (22.5\,{}^m\!/_s)(2.65s) = 59.6m$$

Therefore, Herman must have traveled 59.6 meters horizontally before returning to the Earth.

2.45 Q: A child kicks a ball with an initial velocity of 8.5 meters per second at an angle of 35° with the horizontal, as shown. The ball has an initial vertical velocity of 4.9 meters per second and a total time of flight of 1.0 second. The maximum height reached by the ball is approximately: [Neglect air resistance]

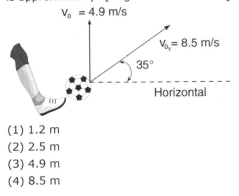

(1) 1.2 m

(2) 2.5 m

(3) 4.9 m

(4) 8.5 m

2.45 A: The maximum height in the air is a vertical motion problem. Start by recognizing that at its maximum height, the ball's vertical velocity will be zero, and it will have been in the air 0.5 seconds.

VERTICAL MOTION TABLE

Variable	Value
v_0	4.9 m/s
v	0 m/s
Δy	FIND
a	-9.8 m/s²
t	0.5 s

$$v^2 = v_0^2 + 2a\Delta y$$

$$\Delta y = \frac{v^2 - v_0^2}{2a}$$

$$\Delta y = \frac{(0\,m/s)^2 - (4.9\,m/s)^2}{2(-9.8\,m/s^2)} = 1.2m$$

The correct answer must be (1) 1.2 meters. Note that you could have solved for the correct answer using any of the kinematic equations containing distance.

2.46 Q: A ball is thrown at an angle of 38° to the horizontal. What happens to the magnitude of the ball's vertical acceleration during the total time interval that the ball is in the air?

(1) It decreases, then increases.

(2) It decreases, then remains the same.

(3) It increases, then decreases.

(4) It remains the same.

2.46 A: (4) It remains the same since the acceleration of any projectile on the surface of Earth is 9.8 m/s² down the entire time the projectile is in the air.

2.47 Q: A golf ball is hit at an angle of 45° above the horizontal. What is the acceleration of the golf ball at the highest point in its trajectory? [Neglect friction.]

(1) 9.8 m/s² upward

(2) 9.8 m/s² downward

(3) 6.9 m/s² horizontal

(4) 0 m/s².

2.47 A: (2) 9.8 m/s² downward.

2.48 Q: A machine launches a tennis ball at an angle of 25° above the horizontal at a speed of 14 meters per second. The ball returns to level ground. Which combination of changes must produce an increase in time of flight of a second launch?

(1) decrease the launch angle and decrease the ball's initial speed

(2) decrease the launch angle and increase the ball's initial speed

(3) increase the launch angle and decrease the ball's initial speed

(4) increase the launch angle and increase the ball's initial speed

2.48 A: (4) will increase the ball's initial vertical velocity and therefore give the ball a larger time of flight.

2.49 Q: A golf ball is given an initial speed of 20 meters per second and returns to level ground. Which launch angle above level ground results in the ball traveling the greatest horizontal distance? [Neglect friction.]

(1) 60°

(2) 45°

(3) 30°

(4) 15°

2.49 A: (2) 45° provides the greatest range for a projectile launched from level ground onto level ground, neglecting friction.

2.50 Q: A 30° incline sits on a 1.1-meter high table. A ball rolls off the incline with a velocity of 2 m/s. How far does the ball travel across the room before reaching the floor?

2.50 A: In order to determine how far the ball travels horizontally, you must first determine how long the ball is in the air. Begin by breaking up its initial velocity into horizontal and vertical components.

$$v_{0_x} = v_0 \cos(\theta) = (2.00 \,{}^m\!/_s) \cos(30°) = 1.73 \,{}^m\!/_s$$

$$v_{0_y} = v_0 \sin(\theta) = (2.00 \,{}^m\!/_s) \sin(30°) = 1.00 \,{}^m\!/_s$$

Determining how long the ball is in the air is a vertical motion problem. If you call the down direction positive, you can set up a vertical motion table as shown below.

VERTICAL MOTION TABLE

Variable	Value
v_0	1 m/s
v	?
Δy	1.1 m
a	9.8 m/s²
t	FIND

To solve this problem, you could solve for time in the kinematic equation: $\Delta y = v_0 t + \frac{1}{2} a t^2$.

However, in doing so, you'll encounter a quadratic equation that will require you to utilize the quadratic formula. This is perfectly solvable, but you can save yourself some time and mathematical complexity if instead you solve for final velocity first, then solve for time.

$$v^2 = v_0^2 + 2a\Delta y$$

$$v^2 = (1\,m/s)^2 + 2(9.8\,m/s^2)(1.1m)$$

$$v^2 = 22.56\,m^2/s^2$$

$$v = \sqrt{22.56\,m^2/s^2} = 4.75\,m/s$$

Now, knowing the ball's final vertical velocity, you can solve for the time the ball is in the air.

$$v = v_0 + at$$

$$t = \frac{v - v_0}{a}$$

$$t = \frac{4.75\,m/s - 1\,m/s}{9.8\,m/s^2} = 0.383s$$

Now, knowing the time the ball is in the air, you can analyze the horizontal motion of the ball to calculate the horizontal distance traveled. Since you're neglecting air resistance, the horizontal acceleration of the ball is zero, therefore the initial velocity is the same as the final velocity.

HORIZONTAL MOTION TABLE

Variable	Value
v_0	1.73 m/s
v	1.73 m/s
Δx	FIND
a	0
t	0.383 s

$$\overline{v} = \frac{\Delta x}{t}$$

$$\Delta x = \overline{v}t = (1.73\,m/s)(0.383s) = 0.66m$$

The ball travels 0.66m across the room horizontally from the edge of the table before striking the ground.

Relative Velocity

You've probably heard the saying "motion is relative." Or perhaps you've heard people speak about Einstein's Theory of General Relativity and Einstein's Theory of Special Relativity. But what is this relativity concept?

In short, the concept of relative motion or relative velocity is all about understanding frame of reference. A frame of reference can be thought of as the state of motion of the observer of some event. For example, if you're sitting on a lawnchair watching a train travel past you from left to right at 50 m/s, you would consider yourself in a stationary frame of reference. From your perspective, you are at rest, and the train is moving. Further, assuming you have tremendous eyesight, you could even watch a glass of water sitting on a table inside the train move from left to right at 50 m/s.

An observer on the train itself, however, sitting beside the table with the glass of water, would view the glass of water as remaining stationary from their frame of reference. Because that observer is moving at 50 m/s, and the glass of water is moving at 50 m/s, the observer on the train sees no motion for the cup of water.

This seems like a simple and obvious example, yet when you take a step back and examine the bigger picture, you quickly find that all motion is relative. Going back to our original scenario, if you're sitting on your lawnchair watching a train go by, you believe you're in a stationary reference frame. The observer on the train looking out the window at you, however, sees you moving from right to left at 50 m/s.

Even more intriguing, an observer outside the Earth's atmosphere traveling with the Earth could use a "magic telescope" to observe you sitting in your lawnchair moving hundreds of meters per second as the Earth rotates about its axis. If this observer were further away from the Earth, he or she would also observe the Earth moving around the sun at speeds approaching 30,000 m/s. If the observer were even further away, they would observe the solar system (with the Earth, and you, on your lawnchair) orbiting the center of the Milky Way Galaxy at speeds approaching 220,000 m/s. And it goes on and on.

According to the laws of physics, there is no way to distinguish between an object at rest and an object moving at a constant velocity in an inertial (non-accelerating) reference frame. This means that there really is no "correct answer" to the question "how fast is the glass of water on the train moving?" You would be correct stating the glass is moving 50 m/s to the right and also correct in stating the glass is stationary. Imagine you're on a very smooth airplane, with all the window shades pulled down. It is physically impossible to determine whether you're flying through the air at a constant 300 m/s or whether you're sitting still on the runway. Even if you peeked out the window, you still couldn't say whether the plane was moving forward at 300 m/s, or the Earth was moving underneath the plane at 300 m/s.

As you observe (pun intended), how fast you are moving depends upon the observer's frame of reference. This is what is meant by the statement "motion is relative." In order to determine an object's velocity, you really need to also state the reference frame (i.e. the train moves 50 m/s with respect to the ground; the glass of water moves 50 m/s with respect to the ground; the glass of water is stationary with respect to the train.)

In most instances, the Earth makes a terrific frame of reference for physics problems. However, there are times when calculating the velocity of an object relative to different reference frames can be useful. Imagine you're in a canoe race, traveling down a river. It could be important to know not only your speed with respect to the flow of the river, but also your speed with respect to the riverbank, and even your speed with respect to your opponent's canoe in the race.

In dealing with these situations, you can state the velocity of an object with respect to its reference frame. For example, the velocity of object A with respect to reference frame C would be written as v_{AC}. Even if you don't know the velocity of object A with respect to C directly, by finding the velocity of object A with respect to some intermediate object B, and the velocity of object B with respect to C, you can combine your velocities using vector addition to obtain:

$$v_{AC} = v_{AB} + v_{BC}$$

This sounds more complicated than it actually is. Let's look at how this is applied in a few examples.

2.51 Q: A train travels at 60 m/s to the east with respect to the ground. A businessman on the train runs at 5 m/s to the west with respect to the train. Find the velocity of the man with respect to the ground.

2.51 A: First determine what information you are given. Calling east the positive direction, you know the velocity of the train with respect to the ground (v_{TG}=60 m/s). You also know the velocity of the man with respect to the train (v_{MT}=-5 m/s). Putting these together, you can find the velocity of the man with respect to the ground.

$$v_{MG} = v_{MT} + v_{TG} = -5\,\tfrac{m}{s} + 60\,\tfrac{m}{s} = 55\,\tfrac{m}{s}$$

2.52 Q: An airplane flies at 250 m/s to the east with respect to the air. The air is moving at 15 m/s to the east with respect to the ground. Find the velocity of the plane with respect to the ground.

2.52 A: Again, start with the information you are given. If you call east positive, the velocity of the plane with respect to the air (v_{PA}) is 250 m/s. The velocity of the air with respect to the ground (v_{AG}) is 15 m/s. Solve for v_{PG}.

$$v_{PG} = v_{PA} + v_{AG} = 250\,^{m}\!/_{s} + 15\,^{m}\!/_{s} = 265\,^{m}\!/_{s}$$

This strategy isn't limited to one-dimensional problems. Treating velocities as vectors, you can use vector addition to solve problems in multiple dimensions.

2.53 Q: The president's airplane, Air Force One, flies at 250 m/s to the east with respect to the air. The air is moving at 35 m/s to the north with respect to the ground. Find the velocity of Air Force One with respect to the ground.

2.53 A: In this case, it's important to realize that both v_{PA} and v_{AG} are two-dimensional vectors. Once again, you can find v_{PG} by vector addition.

$$v_{PG} = v_{PA} + v_{AG}$$

Drawing a diagram can be of tremendous assistance in solving this problem.

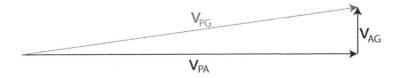

Looking at the diagram, you can easily solve for the magnitude of the velocity of the plane with respect to the ground using the Pythagorean Theorem.

$$v_{PG}^2 = v_{PA}^2 + v_{AG}^2 \qquad v_{PG} = \sqrt{v_{PA}^2 + v_{AG}^2}$$

$$v_{PG} = \sqrt{(250\,^m\!/_s)^2 + (35\,^m\!/_s)^2} = 252\,^m\!/_s$$

You can find the angle of Air Force One using basic trig functions.

$$\tan\theta = \frac{opp}{adj} = \frac{v_{AG}}{v_{PA}} \rightarrow \theta = \tan^{-1}\frac{v_{AG}}{v_{PA}} \rightarrow$$

$$\theta = \tan^{-1}\frac{35\,^m\!/_s}{250\,^m\!/_s} = 8°$$

Therefore, the velocity of Air Force One with respect to the ground is 252 m/s at an angle of 8° north of east.

Chapter 3: Dynamics

"If I have seen further than others, it is by standing upon the shoulders of giants."

— Sir Isaac Newton

Objectives

1. Define mass and inertia and explain the meaning of Newton's 1st Law.
2. Define a force and distinguish between contact forces and field forces.
3. Draw and label a free body diagram showing all forces acting on an object.
4. Determine the resultant of two or more vectors graphically and algebraically.
5. Draw scaled force diagram using a ruler and protractor.
6. Resolve a vector into perpendicular components: both graphically and algebraically.
7. Use vector diagrams to analyze mechanical systems (equilibrium and nonequilibrium).

8. Verify Newton's Second Law for linear motion.
9. Describe how mass and weight are related.
10. Define friction and distinguish between static and kinetic friction.
11. Determine the coefficient of friction for two surfaces.
12. Calculate parallel and perpendicular components of an object's weight to solve ramp problems.
13. Analyze and solve basic Atwood Machine problems using Newton's 2nd Law of Motion.

Now that you've studied kinematics, you should have a pretty good understanding that objects in motion have **kinetic energy**, which is the ability of a moving object to move another object. To change an object's motion, and therefore its kinetic energy, the object must undergo a change in velocity, which is called an **acceleration**. So then, what causes an acceleration? To answer that question, you must study forces and their application.

Dynamics, or the study of forces, was very simply and effectively described by Sir Isaac Newton in 1686 in his masterpiece Principia Mathematica Philosophiae Naturalis. Newton described the relationship between forces and motion using three basic principles. Known as Newton's Laws of Motion, these concepts are still used today in applications ranging from sports science to aeronautical engineering.

Newton's 1st Law of Motion

Newton's 1st Law of Motion, also known as the **law of inertia**, can be summarized as follows:

> "An object at rest will remain at rest, and an object in motion will remain in motion, at constant velocity and in a straight line, unless acted upon by a net force."

This means that unless there is a net (unbalanced) force on an object, an object will continue in its current state of motion with a constant velocity. If this velocity is zero (the object is at rest), the object will continue to remain at rest. If this velocity is not zero, the object will continue to move in a straight line at the same speed. However, if a net (unbalanced) force does act on an object, that object's velocity will be changed (it will accelerate).

This sounds like a simple concept, but it can be quite confusing because it is difficult to observe this in everyday life. People are usually fine with understanding the first part of the law: "an object at rest will remain at rest unless acted upon by a net force." This is easily observable. The donut sitting on your breakfast table this morning didn't spontaneously accelerate up into the sky. Nor did the family cat, Whiskers, lounging sleepily on the couch cushion the previous evening, all of a sudden accelerate sideways off the couch for no apparent reason.

The second part of the law contributes a considerably bigger challenge to the conceptual understanding of this principle. Realizing that "an object in motion will continue in its current state of motion with constant velocity unless acted upon by a net force" isn't easy to observe here on Earth, making this law rather tricky. Almost all objects observed in everyday life that are in motion are being acted upon by a net force - friction. Try this example: take your physics book and give it a good push along the floor. As expected, the book moves for some distance, but rather rapidly slides to a halt. An outside force, friction, has acted upon it. Therefore, from typical observations, it would be easy to think that an object must have a force continually applied upon it to remain in motion. However, this isn't so. If you took the same book out into the far reaches of space, away from any gravitational or frictional forces, and pushed it away, it would continue moving in a straight line at a constant velocity forever and ever, as there are no external forces to change its motion. When the net force on an object is 0, the object is in **static equilibrium**. You'll revisit static equilibrium when discussing Newton's 2nd Law.

The tendency of an object to resist a change in velocity is known as the object's **inertia**. For example, a train has significantly more inertia than a skateboard. It is much harder to change the train's velocity than it is the skateboard's. The measure of an object's inertia is its **mass**. For the purposes of this course, inertia and mass mean the same thing - they are synonymous.

3.1 Q: A 0.50-kilogram cart is rolling at a speed of 0.40 meter per second. If the speed of the cart is doubled, the inertia of the cart is

(1) halved

(2) doubled

(3) quadrupled

(4) unchanged

3.1 A: (4) unchanged. Inertia is another word for mass, and the mass of the cart is constant.

3.2 Q: Which object has the greatest inertia?

(1) a falling leaf

(2) a softball in flight

(3) a seated high school student

(4) a rising helium-filled toy balloon

3.2 A: (3) a seated high school student has the greatest inertia (mass).

3.3 Q: Which object has the greatest inertia?

(1) a 5.00-kg mass moving at 10.0 m/s

(2) a 10.0-kg mass moving at 1.00 m/s

(3) a 15.0-kg mass moving at 10.0 m/s

(4) a 20.0-kg mass moving at 1.00 m/s

3.3 A: (4) a 20.0-kg mass has the greatest inertia.

If you recall from the kinematics unit, a change in velocity is known as an acceleration. Therefore, the second part of this law could be re-written to state that an object acted upon by a net force will be accelerated.

But, what exactly is a force? A **force** is a vector quantity describing the push or a pull on an object. Forces are measured in Newtons (N), named after Sir Isaac Newton, of course. A Newton is not a base unit, but is instead a derived unit, equivalent to 1 kg×m/s². Interestingly, the gravitational force on a medium-sized apple is approximately 1 Newton.

You can break forces down into two basic types: contact forces and field forces. Contact forces occur when objects touch each other. Examples of contact forces include pushing a crate (applied

 force), pulling a wagon (tension force), a frictional force slowing down your sled, or even the force of air accelerating a spitwad through a straw. Field forces, also known as non-contact forces, occur at a distance. Examples of field forces include the gravitational force, the magnetic force, and the electrical force between two charged objects.

So, what then is a net force? A **net force** is just the vector sum of all the forces acting on an object. Imagine you and your sister are fighting over the last Christmas gift. You are pulling one end of the gift toward you with a force of 5N. Your sister is pulling the other end toward her (in the opposite direction) with a force of 5N. The net force on the gift, then, would be 0N, therefore there would be no net force. As it turns out, though, you have a passion for Christmas gifts, and now increase your pulling force to 6N. The net force on the gift now is 1N in your direction, therefore the gift would begin to accelerate toward you (yippee!) It can be difficult to keep track of all the forces acting on an object.

Free Body Diagrams

Fortunately, we have a terrific tool for analyzing the forces acting upon objects. This tool is known as a free body diagram. Quite simply, a **free body diagram** is a representation of a single object, or system, with vector arrows showing all the external forces acting on the object. These diagrams make it very easy to identify exactly what the net force is on an object, and they're also quite simple to create:

1. Isolate the object of interest. Draw the object as a point par-

ticle representing the same mass.

2. Sketch and label each of the external forces acting on the object.
3. Choose a coordinate system, with the direction of motion as one of the positive coordinate axes.
4. If all forces do not line up with your axes, resolve those forces into components using trigonometry (note that the formulas below only work if the angle is measured from the horizontal).

$$A_x = A\cos(\theta)$$
$$A_y = A\sin(\theta)$$

5. Redraw your free body diagram, replacing forces that don't overlap the coordinates axes with their components.

As an example, picture a glass of soda sitting on the dining room table. You can represent the glass of soda in the diagram as a single dot. Then, represent each of the vector forces acting on the soda by drawing arrows and labeling them. In this case, you can start by recognizing the force of gravity on the soda, known more commonly as the soda's **weight**. Although you could label this force as F$_{grav}$, or W, get in the habit right now of writing the force of gravity on an object as mg. You can do this because the force of gravity on an object is equal to the object's mass times the acceleration due to gravity, g.

Of course, since the soda isn't accelerating, there must be another force acting on the soda to balance out the weight. This force, the force of the table pushing up on the soda, is known as the **normal force** (F$_N$). In physics, the normal force refers to a force perpendicular to a surface (normal in this case meaning perpendicular). The force of gravity on the soda must exactly match the normal force on the soda, although they are in opposite directions, therefore there is no net force on the soda. The free body diagram for this situation could be drawn as shown at left.

3.4 Q: Which diagram represents a box in equilibrium?

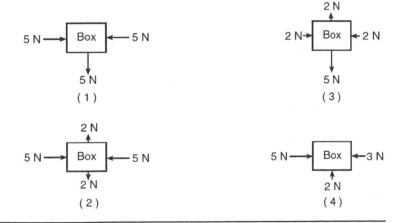

3.4 A: (2) all forces are balanced for a net force of zero.

3.5 Q: If the sum of all the forces acting on a moving object is zero, the object will

(1) slow down and stop

(2) change the direction of its motion

(3) accelerate uniformly

(4) continue moving with constant velocity

3.5 A: (4) continue moving with constant velocity per Newton's 1st Law.

Newton's 2nd Law of Motion

Newton's 2nd Law of Motion may be the most important principle in all of modern-day physics because it explains exactly how an object's velocity is changed by a net force. In words, Newton's 2nd Law states that "the acceleration of an object is directly proportional to the net force applied, and inversely proportional to the object's mass." In equation form:

$$a = \frac{F_{net}}{m}$$

It's important to remember that both force and acceleration are vectors. Therefore, the direction of the acceleration, or the change in velocity, will be in the same direction as the net force. You can also look at this equation from the opposite perspective. A net force applied to an object changes an object's velocity (produces an acceleration), and is frequently written as:

$$F_{net} = ma$$

You can analyze many situations involving both balanced and unbalanced forces on an object using the same basic steps.

1.   Draw a free body diagram.
2.   For any forces that don't line up with the x- or y-axes, break those forces up into components that do lie on the x- or y-axis.
3.   Write expressions for the net force in x- and y- directions. Set the net force equal to ma, since Newton's 2nd Law tells us that F=ma.
4.   Solve the resulting equations.

Let's take a look and see how these steps can be applied to a sample problem.

3.6 Q: A force of 25 newtons east and a force of 25 newtons west act concurrently on a 5-kilogram cart. Find the acceleration of the cart.

(1) 1.0 m/s² west

(2) 0.20 m/s² east

(3) 5.0 m/s² east

(4) 0 m/s²

3.6 A: Step 1: Draw a free-body diagram (FBD).

Step 2: All forces line up with x-axis. Define east as positive.

Step 3: $F_{net} = 25N - 25N = ma$

Step 4: $0 = ma$

$a = 0$

Correct answer must be (4) 0 m/s².

Of course, everything you've already learned about kinematics still applies, and can be applied to dynamics problems as well.

3.7 Q: A 0.15-kilogram baseball moving at 20 m/s is stopped by a catcher in 0.010 seconds. The average force stopping the ball is

(1) 3.0×10⁻² N

(2) 3.0×10⁰ N

(3) 3.0×10¹ N

(4) 3.0×10² N

3.7 A: First write down what information is given and what we're asked to find. Define the initial direction of the baseball as positive.

Given: Find:

$m = 0.15 kg$ F

$v_0 = 20 \, ^m\!/_s$

$v = 0 \, ^m\!/_s$

$t = 0.010 s$

Use kinematics to find acceleration.

$$a = \frac{\Delta v}{t} = \frac{v - v_0}{t} \rightarrow$$

$$a = \frac{0 \, ^m\!/_s - 20 \, ^m\!/_s}{0.010 s} = -2000 \, ^m\!/_{s^2}$$

The negative acceleration indicates the acceleration is in the direction opposite that of the initial velocity of the baseball. Now that you know acceleration, you can solve for force using Newton's 2nd Law.

$$F_{net} = ma = (0.15)(-2000 \ ^m/_{s^2}) = -300 N$$

The correct answer must be (4), 300 newtons. The negative sign in our answer indicates that the force applied is opposite the direction of the baseball's initial velocity.

3.8 Q: Two forces, F_1 and F_2, are applied to a block on a frictionless, horizontal surface as shown below.

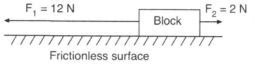

If the magnitude of the block's acceleration is 2.0 meters per second², what is the mass of the block?

(1) 1 kg

(2) 5 kg

(3) 6 kg

(4) 7 kg

3.8 A: Define left as the positive direction.

12 N ⬅—————⬤———➡ 2N

$$F_{net} = 12N - 2N = 10N = ma$$

$$m = \frac{F_{net}}{a} = \frac{10N}{2 \ ^m/_{s^2}} = 5kg$$

(2) 5 kg

3.9 Q: What is the weight of a 2.00-kilogram object on the surface of Earth?

(1) 4.91 N

(2) 2.00 N

(3) 9.81 N

(4) 19.6 N

3.9 A: Weight is the force of gravity on an object. From Newton's 2nd Law, the force of gravity on an object (F_g), is equal to the mass of the object times its acceleration, the acceleration due to gravity (9.81 m/s²), which you can abbreviate as g.

$$F_g = ma$$

$$W = mg$$

$$W = (2kg)(9.8 \; ^m/_{s^2}) = 19.6N$$

(4) 19.6 N is correct.

3.10 Q: A 25-newton horizontal force northward and a 35-newton horizontal force southward act concurrently on a 15-kilogram object on a frictionless surface. What is the magnitude of the object's acceleration?

(1) 0.67 m/s²

(2) 1.7 m/s²

(3) 2.3 m/s²

(4) 4.0 m/s²

3.10 A: (1) 0.67 m/s².

$$a = \frac{F_{net}}{m}$$

$$a = \frac{35N - 25N}{15kg} = 0.67 \; ^m/_{s^2}$$

Static Equilibrium

The special situation in which the net force on an object turns out to be zero, called **static equilibrium**, tells you immediately that the object isn't accelerating. If the object is moving with some velocity, it will remain moving with that exact same velocity. If the object is at rest, it will remain at rest. Sounds familiar, doesn't it? This is a restatement of Newton's 1st Law of Motion, the Law of Inertia. So in reality, Newton's 1st Law is just a special case of Newton's 2nd Law, describing static equilibrium conditions! Consider the situation of a tug-of-war... if both participants are pulling with tremendous force, but the force is balanced, there is no acceleration -- a great example of static equilibrium.

Static equilibrium conditions are so widespread that knowing how to explore and analyze these conditions is a key stepping stone to understanding more complex situations.

One common analysis question involves finding the equilibrant force given a free body diagram of an object. The
equilibrant is a single force vector that you add to the unbalanced forces on an object in order to bring it into static
equilibrium. For example, if you are given a force vector of
10N north and 10N east, and asked to find the equilibrant,
you're really being asked to find a force that will offset the
two given forces, bringing the object into static equilibrium.

To find the equilibrant, you must first find the net force
being applied to the object. To do this, apply your vector math and add up the two vectors by first lining them
up tip to tail, then drawing a straight line from the starting point of the first vector to the ending point of the last
vector. The magnitude of this vector can be found from
the Pythagorean Theorem.

Finally, to find the equilibrant vector, add a single vector to the diagram that will give a net force of zero. If your total net force
is currently 14N northeast, then the vector that should bring
this back into equilibrium, the equilibrant, must be the opposite of 14N northeast, or a vector with magnitude 14N
to the southwest.

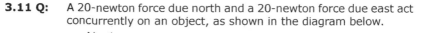

3.11 Q: A 20-newton force due north and a 20-newton force due east act
concurrently on an object, as shown in the diagram below.

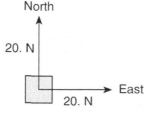

North

20. N

East

20. N

The additional force necessary to bring the object into a state of
equilibrium is

(1) 20 N northeast

(2) 20 N southwest

(3) 28 N northeast

(4) 28 N southwest

3.11 A: (4) The resultant vector is 28 newtons northeast, so its equilibrant
must be 28 newtons southwest

Another common analysis question involves asking whether three vectors could be arranged to provide a static equilibrium situation.

3.12 Q: A 3-newton force and a 4-newton force are acting concurrently on a point. Which force could not produce equilibrium with these two forces?

(1) 1 N

(2) 7 N

(3) 9 N

(4) 4 N

3.12 A: (3) A 9-newton force could not produce equilibrium with a 3-newton and a 4-newton force.

3.13 Q: A net force of 10 newtons accelerates an object at 5.0 meters per second². What net force would be required to accelerate the same object at 1.0 meter per second²?

(1) 1.0 N

(2) 2.0 N

(3) 5.0 N

(4) 50 N

3.13 A: Strategy: First, solve for the mass of the object.

$$F_{net} = ma$$

$$m = \frac{F_{net}}{a} = \frac{10N}{5\,^m/_{s^2}} = 2kg$$

Next, use Newton's 2nd Law to determine the force required to accelerate the object at 1 m/s².

$$F_{net} = ma = (2kg)(1\,^m/_{s^2}) = 2N$$

The correct answer is (2) 2.0 N.

3.14 Q: A 1.0-newton metal disk rests on an index card that is balanced on top of a glass. What is the net force acting on the disk?

(1) 1 N

(2) 2 N

(3) 0 N

(4) 9.8 N

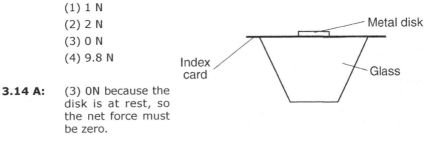

3.14 A: (3) 0N because the disk is at rest, so the net force must be zero.

3.15 Q: A 1200-kilogram space vehicle travels at 4.8 meters per second along the level surface of Mars. If the magnitude of the gravitational field strength on the surface of Mars is 3.7 newtons per kilogram, the magnitude of the normal force acting on the vehicle is

(1) 320 N

(2) 930 N

(3) 4400 N

(4) 5800 N

3.15 A: If the gravitational field strength is 3.7 N/kg, and the space vehicle weighs 1200 kg, the gravitational force on the space vehicle must be 1200kg × (3.7 N/kg) = 4440N. If there's a downward force of 4440N due to gravity, the normal force must be equal and opposite, or 4440N upward, therefore the best answer is (3) 4400N.

3.16 Q: Which body is in equilibrium?

(1) a satellite orbiting Earth in a circular orbit

(2) a ball falling freely toward the surface of Earth

(3) a car moving with constant speed along a straight, level road

(4) a projectile at the highest point in its trajectory

3.16 A: (3) a car moving with constant speed.

Newton's 3rd Law

Newton's 3rd Law of Motion, commonly referred to as the Law of Action and Reaction, describes the phenomena by which all forces come in pairs. If Object 1 exerts a force on Object 2, then Object 2 must exert a force back on Object 1. Moreover, the force of Object 1 on Object 2 is equal in magnitude, or size, but opposite in direction to the force of Object 2 on Object 1. Written mathematically:

$$\vec{F}_{1on2} = -\vec{F}_{2on1}$$

This has many implications, many of which aren't immediately obvious. For example, if you punch the wall with your fist with a force of 100N, the wall imparts a force back on your fist of 100N (which is why it hurts!). Or try this. Push on the corner of your desk with your palm for a few seconds. Now look at your palm... see the indentation? That's because the corner of the desk pushed back on your palm.

Although this law surrounds your actions every-day, often times you may not even realize its effects. To run forward, a cat pushes with its legs backward on the ground, and the ground pushes the cat forward. How do you swim? If you want to swim forwards, which way do you push on the water? Backwards, that's right. As you push backwards on the water, the reactionary force, the water pushing you, propels you forward. How do you jump up in the air? You push down on the ground, and it's the reactionary force of the ground pushing on you that accelerates you skyward!

As you can see, then, forces always come in pairs. These pairs are known as **action-reaction pairs**. What are the action-reaction force pairs for a girl kicking a soccer ball? The girl's foot applies a force on the ball, and the ball applies an equal and opposite force on the girl's foot.

How does a rocket ship maneuver in space? The rocket propels hot expanding gas particles outward, so the gas particles in return push the rocket forward. Newton's 3rd Law even applies to gravity. The Earth exerts a gravitational force on you (downward). You, therefore, must apply a gravitational force upward on the Earth!

3.17 Q: Earth's mass is approximately 81 times the mass of the Moon. If Earth exerts a gravitational force of magnitude F on the Moon, the magnitude of the gravitational force of the Moon on Earth is

(1) F

(2) F/81

(3) 9F

(4) 81F

3.17 A: (1) The force Earth exerts on the Moon is the same in magnitude and opposite in direction of the force the Moon exerts on Earth.

3.18 Q: A 400-newton girl standing on a dock exerts a force of 100 newtons on a 10,000-newton sailboat as she pushes it away from the dock. How much force does the sailboat exert on the girl?

(1) 25 N

(2) 100 N

(3) 400 N

(4) 10,000 N

3.18 A: (2) The force the girl exerts on the sailboat is the same in magnitude and opposite in direction of the force the sailboat exerts on the girl.

3.19 Q: A carpenter hits a nail with a hammer. Compared to the magnitude of the force the hammer exerts on the nail, the magnitude of the force the nail exerts on the hammer during contact is

(1) less

(2) greater

(3) the same

3.19 A: (3) the same per Newton's 3rd Law.

Friction

Up until this point, it's been convenient to ignore one of the most useful and most troublesome forces in everyday life... a force that has tremendous application in transportation, machinery, and all parts of mechanics, yet people spend tremendous amounts of effort and money each day fighting it. This force, **friction**, is a force that opposes motion.

3.20 Q: A projectile launched at an angle of 45° above the horizontal travels through the air. Compared to the projectile's theoretical path with no air friction, the actual trajectory of the projectile with air friction is

(1) lower and shorter

(2) lower and longer

(3) higher and shorter

(4) higher and longer

3.20 A: (1) lower and shorter. Friction opposes motion.

3.21 Q: A box is pushed toward the right across a classroom floor. The force of friction on the box is directed toward the

(1) left

(2) right

(3) ceiling

(4) floor

3.21 A: (1) left. Friction opposes motion.

There are two main types of friction. **Kinetic friction** is a frictional force that opposes motion for an object which is sliding along another surface. **Static friction**, on the other hand, acts on an object that isn't sliding. If you push on your textbook, but not so hard that it slides along your desk, static friction is opposing your applied force on the book, leaving the book in static equilibrium.

The magnitude of the frictional force depends upon two factors:

1. The nature of the surfaces in contact.
2. The normal force acting on the object (F_N).

The ratio of the frictional force and the normal force provides the **coefficient of friction** (μ), a proportionality constant that is specific to the two materials in contact.

You can look up the coefficient of friction for various surfaces. Make sure you choose the appropriate coefficient. Use the static coefficient (μ_s) for objects which are not sliding, and the kinetic coefficient (μ_k) for objects which are sliding.

Approximate Coefficients of Friction		
	Kinetic	Static
Rubber on concrete (dry)	0.68	0.90
Rubber on concrete (wet)	0.58	
Rubber on asphalt (dry)	0.67	0.85
Rubber on asphalt (wet)	0.53	
Rubber on ice	0.15	
Waxed ski on snow	0.05	0.14
Wood on wood	0.30	0.42
Steel on steel	0.57	0.74
Copper on steel	0.36	0.53
Teflon on Teflon	0.04	

Which coefficient would you use for a sled sliding down a snowy hill? The kinetic coefficient of friction, of course. How about a refrigerator on your linoleum floor that is at rest and you want to start in motion? That would be the static coefficient of friction. Let's try a harder one... A car drives with its tires rolling freely. Is the friction between the tires and the road static or kinetic? Static. The tires are in constant contact with the road, much like walking. If the car was skidding, however, and the tires were locked, you would look at kinetic friction. Let's take a look at a sample problem:

3.22 Q: A car's performance is tested on various horizontal road surfaces. The brakes are applied, causing the rubber tires of the car to slide along the road without rolling. The tires encounter the greatest force of friction to stop the car on

(1) dry concrete

(2) dry asphalt

(3) wet concrete

(4) wet asphalt

3.22 A: To obtain the greatest force of friction (F_f), you'll need the greatest coefficient of friction (μ). Use the kinetic coefficient of friction (μ_k) since the tires are sliding. From the Approximate Coefficients of Friction table, the highest kinetic coefficient of friction for rubber comes from rubber on dry concrete. Answer: (1).

3.23 Q: The diagram below shows a block sliding down a plane inclined at angle θ with the horizontal.

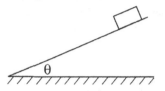

As angle θ is increased, the coefficient of kinetic friction between the bottom surface of the block and the surface of the incline will

(1) decrease

(2) increase

(3) remain the same

3.23 A: (3) remain the same. Coefficient of friction depends only upon the materials in contact.

The normal force always acts perpendicular to a surface, and comes from the interaction between atoms that act to maintain its shape. In many cases, it can be thought of as the elastic force trying to keep a flat surface flat (instead of bowed). You can use the normal force to calculate the magnitude of the frictional force.

The force of friction, depending only upon the nature of the surfaces in contact (μ) and the magnitude of the normal force (F_N), can be determined using the formula:

$$F_f = \mu F_N$$

Solving problems involving friction requires us application of the same basic principles you've been learning about throughout the dynamics unit... drawing a free body diagram, applying Newton's 2nd Law along the x- and/ or y-axes, and solving for any unknowns. The only new skill is drawing the frictional force on the free body diagram, and using the relationship between the force of friction and the normal force to solve for the unknowns. Let's take a look at another sample problem:

3.24 Q: The diagram below shows a 4.0-kilogram object accelerating at 10 meters per second² on a rough horizontal surface.

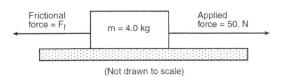

(Not drawn to scale)

What is the magnitude of the frictional force F_f acting on the object?

(1) 5.0 N

(2) 10 N

(3) 20 N

(4) 40 N

3.24 A: Define to the right as the positive direction.

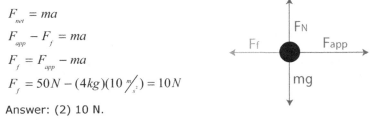

$$F_{net} = ma$$

$$F_{app} - F_f = ma$$

$$F_f = F_{app} - ma$$

$$F_f = 50N - (4kg)(10\,{}^m\!/_{s^2}) = 10N$$

Answer: (2) 10 N.

Let's take a look at a more involved problem, tying together free body diagrams, Newton's 2nd Law, and the coefficient of friction:

3.25 Q: An ice skater applies a horizontal force to a 20-kilogram block on frictionless, level ice, causing the block to accelerate uniformly at 1.4 m/s² to the right. After the skater stops pushing the block, it slides onto a region of ice that is covered by a thin layer of sand. The coefficient of kinetic friction between the block and the sand-covered ice is 0.28. Calculate the magnitude of the force applied to the block by the skater.

3.25 A: Define right as the positive direction.

$$F_{net} = ma$$

$$F_{net} = (20kg)(1.4\,{}^m\!/_{s^2}) = 28N$$

3.26 Q: Referring to the previous problem, determine the magnitude of the normal force acting on the block.

3.26 A: $F_{net_y} = ma_y$

$F_{net_y} = F_N - mg = 0$

$F_N = mg = (20kg)(9.8 \,{}^m\!/_{s^2}) = 196N$

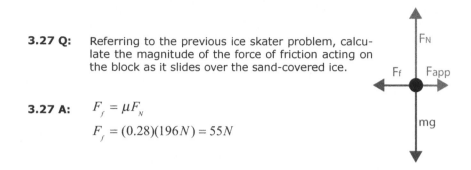

3.27 Q: Referring to the previous ice skater problem, calculate the magnitude of the force of friction acting on the block as it slides over the sand-covered ice.

3.27 A: $F_f = \mu F_N$

$F_f = (0.28)(196N) = 55N$

These same steps can be used in many different ways in many different problems, but the same basic problem solving methodology still works... draw a free body diagram, apply Newton's 2nd Law, utilize the friction formula if necessary, and solve!

3.28 Q: A horizontal force of 8.0 newtons is used to pull a 20-newton wooden box moving toward the right along a horizontal, wood surface, as shown.

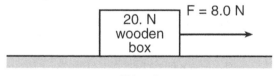

Calculate the magnitude of the frictional force acting on the box.

3.28 A: Recognize that the box has a weight of 20N, therefore its normal force must be 20N since it is not accelerating vertically.

$F_f = \mu F_N = (0.30)(20N) = 6N$

3.29 Q: Referring to the problem above, determine the magnitude of the net force acting on the box.

3.29 A: $F_{net} = F_{app} - F_f = 8N - 6N = 2N$

3.30 Q: Referring to the problem of 3.28, determine the mass of the box.

3.30 A: $mg = 20N$

$$m = \frac{20N}{g} = \frac{20N}{9.8\,{}^{m}\!/_{s^2}} = 2.04kg$$

3.31 Q: Referring to the problem of 3.28, calculate the magnitude of the acceleration of the box.

3.31 A: $a = \frac{F_{net}}{m} = \frac{2N}{2.04kg} = 0.98\,{}^{m}\!/_{s^2}$

3.32 Q: Compared to the force needed to start sliding a crate across a rough level floor, the force needed to keep it sliding once it is moving is

(1) less

(2) greater

(3) the same

3.32 A: (1) less, since kinetic friction is less than static friction.

3.33 Q: An airplane is moving with a constant velocity in level flight. Compare the magnitude of the forward force provided by the en- gines to the magnitude of the backward frictional drag force.

3.33 A: The forces must be the same since the plane is moving with constant velocity.

3.34 Q: When a 12-newton horizontal force is applied to a box on a hori- zontal tabletop, the box remains at rest. The force of static fric- tion acting on the box is

(1) 0 N

(2) between 0 N and 12 N

(3) 12 N

(4) greater than 12 N

3.34 A: (3) 12 N. Because the box is at rest in static equilibrium, all forces on it must be balanced, therefore the force of static friction must be 12N.

Ramps and Inclines

Now that you've developed an understanding of Newton's Laws of Motion, free body diagrams, friction, and forces on flat surfaces, you can extend these tools to situations on ramps, or inclined surfaces. The key to understanding these situations is creating an accurate free body diagram after choosing convenient x- and y-axes. Problem-solving steps are consistent with those developed for Newton's 2nd Law.

Take the example of a box on a ramp inclined at an angle of θ with respect to the horizontal. You can draw a basic free body diagram for this situation, with the force of gravity pulling the box straight down, the normal force perpendicular out of the ramp, and friction opposing motion (in this case pointing up the ramp).

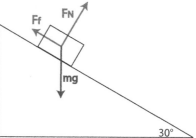

Once the forces acting on the box have been identified, you must be clever about our choice of x-axis and y-axis directions. Much like analyzing free falling objects and projectiles, if you set the positive x-axis in the direction of initial motion (or the direction the object wants to move if it is not currently moving), the y-axis must lie perpendicular to the ramp's surface (parallel to the normal force). Now, you can re-draw the free body diagram, this time superimposing it on the new axes.

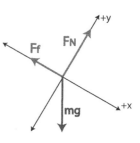

Unfortunately, the force of gravity on the box, *mg*, doesn't lie along one of the axes. Therefore, it must be broken up into components which do lie along the x- and y-axes in order to simplify the mathematical analysis. To do this, you can use geometry to break the weight down into a component parallel with the axis of motion (mg∥) and a component perpendicular to the x-axis (mg⊥) using trigonometry:

$$mg_{\parallel} = mg\sin(\theta)$$

$$mg_{\perp} = mg\cos(\theta)$$

You can now re-draw the free body diagram, replacing *mg* with its components. All the forces line up with the axes, making it straightforward to write Newton's 2nd Law Equations (F_{NETx} and F_{NETy}) and continue with your standard problem-solving strategy.

In the example shown with the modified free body diagram, you could write the Newton's 2nd Law Equations for both the x- and y-directions as follows:

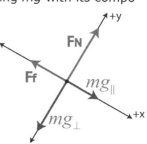

$$F_{net_x} = mg_{\parallel} - F_f = mg\sin(\theta) - F_f = ma_x$$

$$F_{net_y} = F_N - mg_{\perp} = F_N - mg\cos(\theta) = 0$$

From this point, the problem becomes an exercise in algebra. If you need to tie the two equations together to eliminate a variable, don't forget the equation for the force of friction.

3.35 Q: Three forces act on a box on an inclined plane as shown in the diagram below. [Vectors are not drawn to scale.]

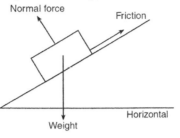

If the box is at rest, the net force acting on it is equal to

(1) the weight

(2) the normal force

(3) friction

(4) zero

3.35 A: (4) zero. If the box is at rest, the acceleration must be zero, therefore the net force must be zero.

3.36 Q: A 5-kg mass is held at rest on a frictionless 30° incline by force F. What is the magnitude of F?

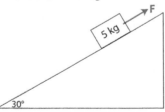

3.36 A: Start by identifying the forces on the box and making a free body diagram.

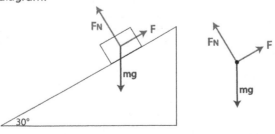

Break up the weight of the box into components parallel to and perpendicular to the ramp, and re-draw the free body diagram using the components of the box's weight.:

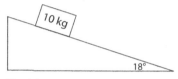

$$mg_{\parallel} = mg\sin(\theta)$$

$$mg_{\perp} = mg\cos(\theta)$$

Finally, use Newton's 2nd Law in the x-direction to solve for the force F.

$$F_{net} = F - mg_{\parallel} = F - mg\sin(\theta) = 0$$

$$F = mg\sin(\theta)$$

$$F = (5kg)(9.8\,^m\!/_{s^2})\sin(30°) = 24.5N$$

3.37 Q: A 10-kg box slides down a frictionless 18° ramp. Find the acceleration of the box, and the time it takes the box to slide 2 meters down the ramp.

3.37 A: Start by identifying the forces on the box and making a free body diagram.

Break up the weight of the box into components parallel to and perpendicular to the ramp.

$$mg_{\parallel} = mg\sin(\theta)$$

$$mg_{\perp} = mg\cos(\theta)$$

Use Newton's 2nd Law to find the acceleration.

$$F_{net} = mg_{\parallel} = mg\sin(\theta) = ma$$

$$a = g\sin(\theta) = (9.8\,^m\!/_{s^2})(\sin(18°)) = 3.03\,^m\!/_{s^2}$$

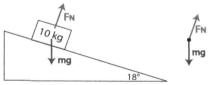

Finally, use the acceleration to solve for the time it takes the box to travel 2m down the ramp using kinematic equations.

$$\Delta x = v_0 t + \tfrac{1}{2}at^2 = (0) + \tfrac{1}{2}at^2$$

$$t = \sqrt{\frac{2\Delta x}{a}} = \sqrt{\frac{2(2m)}{3.03\,^m\!/_{s^2}}} = 1.15s$$

3.38 Q: The diagram below shows a 1.0 × 10⁵-newton truck at rest on a hill that makes an angle of 8.0° with the horizontal.

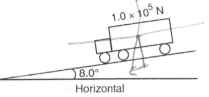

1.0×10^5 N

8.0°

Horizontal

What is the component of the truck's weight parallel to the hill?

(1) 1.4 × 10³-newton
(2) 1.0 × 10⁴-newton
(3) 1.4 × 10⁴-newton
(4) 9.9 × 10⁴-newton

$F_{net} = mg \sin \theta$
$(1.0 \times 10^5)(\sin(8)) \ N$

3.38 A: (3) 1.4 × 10⁴-newton

$$mg_{\parallel} = mg\sin(\theta) = (1.0 \times 10^5 \, N)(\sin(8°)) = 1.4 \times 10^4 \, N$$

3.39 Q: A block weighing 10 newtons is on a ramp inclined at 30° to the horizontal. A 3-newton force of friction, F_f, acts on the block as it is pulled up the ramp at constant velocity with force F, which is parallel to the ramp, as shown in the diagram below.

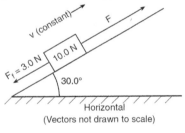

v (constant)

F

$F_f = 3.0 \, N$ 10.0 N

30.0°

Horizontal

(Vectors not drawn to scale)

What is the magnitude of force F?

(1) 7 N
(2) 8 N
(3) 10 N
(4) 13 N

3.39 A: (2) Draw FBD and break weight of the box up into components.

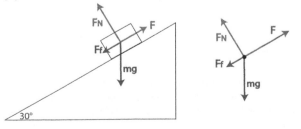

F_N F

F_f

mg

F_N F

F_f

mg

30°

Break up the weight of the box into components parallel to and perpendicular to the ramp, then use Newton's second law, realizing that the forces must be balanced (net force is zero) since the box is moving at constant velocity.

$$F_{net} = F - F_f - mg\sin(\theta) = 0$$

$$F = F_f + mg\sin(\theta) = 3N + 10N \times \sin(30°) = 8N$$

3.40 Q: Which vector diagram best represents a cart slowing down as it travels to the right on a horizontal surface?

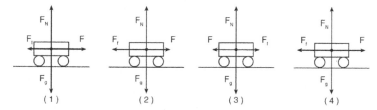

3.40 A: (2) Vertical forces are balanced, net force horizontally is in opposite direction of motion to create a negative acceleration and slow the cart down.

3.41 Q: The diagram represents a block at rest on an incline. Which diagram best represents the forces acting on the block? (F_f = frictional force, F_N = normal force, and F_w = weight.)

3.41 A: Correct Answer is (4).

Atwood Machines

An Atwood Machine is a basic physics laboratory device often used to demonstrate basic principles of dynamics and acceleration. The machine typically involves a pulley, a string, and a system of masses. Keys to solving Atwood

Machine problems are recognizing that the force transmitted by a string or rope, known as tension, is constant throughout the string, and choosing a consistent direction as positive. Let's walk through an example to demonstrate.

3.42 Q: Two masses, m_1 and m_2, are hanging by a massless string from a frictionless pulley. If m_1 is greater than m_2, determine the acceleration of the two masses when released from rest.

3.42 A: First, identify a direction as positive. Since you can easily observe that m_1 will accelerate downward and m_2 will accelerate upward, since $m_1 > m_2$, call the direction of motion around the pulley and down toward m_1 the positive y direction. Then, you can create free body diagrams for both object m_1 and m_2, as shown below:

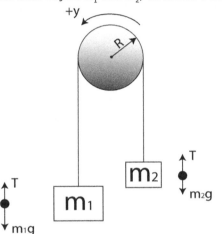

Using this diagram, write Newton's 2nd Law equations for both objects, taking care to note the positive y direction:

$$m_1 g - T = m_1 a \quad (1)$$
$$T - m_2 g = m_2 a \quad (2)$$

Next, combine the equations and eliminate T by solving for T in equation (2) and substituting in for T in equation (1).

$$T - m_2 g = m_2 a \qquad (2)$$
$$T = m_2 g + m_2 a \qquad (2b)$$
$$m_1 g - m_2 g - m_2 a = m_1 a \qquad (1+2b)$$

Finally, solve for the acceleration of the system.

$$m_1 g - m_2 g - m_2 a = m_1 a \quad (1+2b)$$

$$m_1 g - m_2 g = m_1 a + m_2 a$$

$$g(m_1 - m_2) = a(m_1 + m_2)$$

$$a = g\frac{\left(m_1 - m_2\right)}{\left(m_1 + m_2\right)}$$

Alternately, you could treat both masses as part of the same system.

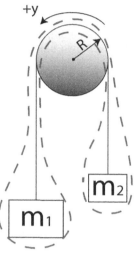

Drawing a dashed line around the system, you can directly write an appropriate Newton's 2nd Law equation for the entire system.

$$m_1 g - m_2 g = (m_1 + m_2)a$$

$$a = g\frac{\left(m_1 - m_2\right)}{\left(m_1 + m_2\right)}$$

Note that if the string and pulley were not massless, this problem would get considerably more involved.

3.43 Q: Two masses are hung from a frictionless pulley by a massless spring. If m_1 is 5 kg, and m_2 is 7 kg, how far will m_2 fall in 2 seconds if released from rest?

3.43 A: This is both a dynamics and a kinematics problem. To solve this problem, you must first find the acceleration of m_2. Once you know the acceleration, you can use kinematics to determine how far m_2 falls in the given time interval.

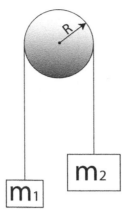

In this problem, it is obvious that m_2 will accelerate toward the ground while m_1 accelerates upward, so choose the positive y-direction accordingly.

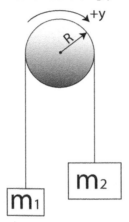

Looking at this from the systems approach, you can write the Newton's 2nd Law Equation as follows:

$$F_{NET_y} = m_2g - m_1g = (m_1 + m_2)a$$

$$a = g\frac{\left(m_2 - m_1\right)}{\left(m_2 + m_1\right)}$$

$$a = \left(9.8 \; ^m\!/_{s^2}\right) \times \frac{\left(7kg - 5kg\right)}{\left(7kg + 5kg\right)}$$

$$a = 1.63 \; ^m\!/_{s^2}$$

Now, knowing the acceleration of the system, find how far m_2 falls in two seconds using the kinematic equations.

VERTICAL MOTION TABLE

Variable	Value
v_0	0 m/s
v	?
Δy	FIND
a	1.63 m/s²
t	2 s

With the given and find information clearly defined, you can utilize the kinematic equations to determine how far the mass fell in two seconds.

$$\Delta y = v_0 t + \tfrac{1}{2} a t^2$$

$$\Delta y = 0 + \tfrac{1}{2}(1.63 \, ^m\!/_{s^2})(2s)^2$$

$$\Delta y = 3.27 m$$

Chapter 4: Momentum

"Sliding headfirst is the safest way to get to the next base, I think, and the fastest. You don't lose your momentum, and there's one more important reason I slide headfirst, it gets my picture in the paper."

— *Pete Rose*

Objectives

1. Define and calculate the momentum of an object.
2. Determine the impulse given to an object.
3. Use impulse to solve a variety of problems.
4. Interpret and use force vs. time graphs.
5. Apply conservation of momentum to solve a variety of problems.
6. Distinguish between elastic and inelastic collisions.
7. Calculate the center of mass for a system of point particles.

You've explored motion in some depth, specifically trying to relate what you know about motion back to kinetic energy. Recall the definition of kinetic energy as the ability or capacity of a moving object to move another object. The key characteristics of kinetic energy, mass and velocity, can be observed from the equation:

$$KE = \tfrac{1}{2}mv^2$$

There's more to the story, however. Moving objects may cause other objects to move, but these interactions haven't been explored yet. To learn more about how one object causes another to move, you need to learn about collisions, and collisions are all about **momentum**.

Defining Momentum

Assume there's a car speeding toward you, out of control without its brakes, at a speed of 27 m/s (60 mph). Can you stop it by standing in front of it and holding out your hand? Why not?

Unless you're Superman, you probably don't want to try stopping a moving car by holding out your hand. It's too big, and it's moving way too fast. Attempting such a feat would result in a number of physics demonstrations upon your body, all of which would hurt.

You can't stop the car because it has too much momentum. **Momentum** is a vector quantity, given the symbol p, which measures how hard it is to stop a moving object. Of course, larger objects have more momentum than smaller objects, and faster objects have more momentum than slower objects. You can therefore calculate momentum using the equation:

$$p = mv$$

Momentum is the product of an object's mass times its velocity, and its units must be the same as the units of mass [kg] times velocity [m/s], therefore the units of momentum must be [kg·m/s], which can also be written as a Newton-second [N·s].

4.01 Q: Two trains, Big Red and Little Blue, have the same velocity. Big Red, however, has twice the mass of Little Blue. Compare their momenta.

4.01 A: Because Big Red has twice the mass of Little Blue, Big Red must have twice the momentum of Little Blue.

4.02 Q: The magnitude of the momentum of an object is 64 kilogram-meters per second. If the velocity of the object is doubled, the magnitude of the momentum of the object will be

(1) 32 kg·m/s

(2) 64 kg·m/s

(3) 128 kg·m/s

(4) 256 kg·m/s

4.02 A: (3) if velocity is doubled, momentum is doubled.

Because momentum is a vector, the direction of the momentum vector is the same as the direction of the object's velocity.

4.03 Q: An Aichi D3A bomber, with a mass of 3600 kg, departs from its aircraft carrier with a velocity of 85 m/s due east. What is the jet's momentum?

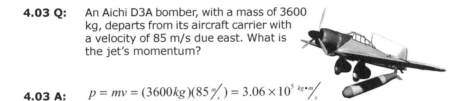

4.03 A: $p = mv = (3600kg)(85 \frac{m}{s}) = 3.06 \times 10^5 \ \frac{kg \bullet m}{s}$

Now, assume the bomber drops its payload and has burned up most of its fuel as it continues its journey east to its destination air field.

4.04 Q: If the bomber's new mass is 3,000 kg, and due to its reduced weight the pilot increases the cruising speed to 120 m/s, what is the bomber's new momentum?

4.04 A: $p = mv = (3000kg)(120 \frac{m}{s}) = 3.60 \times 10^5 \ \frac{kg \bullet m}{s}$

4.05 Q: Cart A has a mass of 2 kilograms and a speed of 3 meters per second. Cart B has a mass of 3 kilograms and a speed of 2 meters per second. Compared to the inertia and magnitude of momentum of cart A, cart B has

(1) the same inertia and a smaller magnitude of momentum

(2) the same inertia and the same magnitude of momentum

(3) greater inertia and a smaller magnitude of momentum

(4) greater inertia and the same magnitude of momentum

4.05 A: (4) greater inertia and the same magnitude of momentum.

Impulse

As you can see, momentum can change, and a change in momentum is known as an **impulse**. In physics, the vector quantity impulse is represented by a capital J, and since it's a change in momentum, its units are the same as those for momentum, [kg·m/s], and can also be written as a Newton-second [N·s].

$$J = \Delta p$$

4.06 Q: Assume the D3A bomber, which had a momentum of 3.6×10^5 kg·m/s, comes to a halt on the ground. What impulse is applied?

4.06 A: Define east as the positive direction:

$$J = \Delta p = p - p_0 = 0 - 3.6 \times 10^5 \; {}^{kg \bullet m}\!\!/\!\!{}_s$$
$$J = -3.6 \times 10^5 \; {}^{kg \bullet m}\!\!/\!\!{}_s \; \text{east} = 3.6 \times 10^5 \; {}^{kg \bullet m}\!\!/\!\!{}_s \; \text{west}$$

4.07 Q: Calculate the magnitude of the impulse applied to a 0.75-kilogram cart to change its velocity from 0.50 meter per second east to 2.00 meters per second east.

4.07 A: $J = \Delta p = m\Delta v = (0.75 kg)(1.5 \, {}^{m}\!\!/\!\!{}_s) = 1.1 N \bullet s$

4.08 Q: A 6.0-kilogram block, sliding to the east across a horizontal, frictionless surface with a momentum of 30 kilogram•meters per second, strikes an obstacle. The obstacle exerts an impulse of 10 newton•seconds to the west on the block. The speed of the block after the collision is

(1) 1.7 m/s
(2) 3.3 m/s
(3) 5.0 m/s
(4) 20 m/s

4.08 A: (2) $J = \Delta p = m\Delta v = m(v - v_0) = mv - mv_0$

$$v = \frac{J + mv_0}{m} = \frac{(-10 N \bullet s) + 30 \; {}^{kg \bullet m}\!\!/\!\!{}_s}{6 kg} = 3.3 \, {}^{m}\!\!/\!\!{}_s$$

4.09 Q: Which two quantities can be expressed using the same units?
(1) energy and force
(2) impulse and force
(3) momentum and energy
(4) impulse and momentum

4.09 A: (4) impulse and momentum both have units of kg·m/s.

4.10 Q: A 1000-kilogram car traveling due east at 15 meters per second is hit from behind and receives a forward impulse of 6000 newton-seconds. Determine the magnitude of the car's change in momentum due to this impulse.

4.10 A: Change in momentum is the definition of impulse, therefore the answer must be 6000 newton-seconds.

Impulse-Momentum Theorem

Since momentum is equal to mass times velocity, you can write that p=mv. Since you also know impulse is a change in momentum, impulse can be written as J=Δp. Combining these equations, you find:

$$J = \Delta p = \Delta mv$$

Since the mass of a single object is constant, a change in the product of mass and velocity is equivalent to the product of mass and change in velocity. Specifically:

$$J = \Delta p = m\Delta v$$

A change in velocity is called acceleration. But what causes an acceleration? A force! And does it matter if the force is applied for a very short time or a very long time? Common sense says it does and also tells us that the longer the force is applied, the longer the object will accelerate, and therefore the greater the object's change in momentum!

You can prove this using an old mathematician's trick -- if you multiply the right side of the equation by 1, you of course get the same thing. And if you multiply the right side of the equation by Δt/Δt, which is 1, you still get the same thing. Take a look:

$$J = \Delta p = \frac{m\Delta v\Delta t}{\Delta t}$$

If you look carefully at this equation, you can find a Δv/Δt, which is, by definition, acceleration. By replacing Δv/Δt with acceleration *a* in the equation, you arrive at:

$$J = \Delta p = ma\Delta t$$

One last step... perhaps you can see it already. On the right-hand side of this equation, you have ma∆t. Utilizing Newton's 2nd Law, you can replace the product of mass and acceleration with force F, giving the final form of the equation, oftentimes referred to as the Impulse-Momentum Theorem:

$$J = \Delta p = F\Delta t$$

This equation relates impulse to change in momentum to force applied over a time interval. For the same change in momentum, force can vary by changing the time over which it is applied. Great examples include airbags in cars, boxers rolling with punches, skydivers bending their knees upon landing, etc. To summarize, when an unbalanced force acts on an object for a period of time, a change in momentum is produced, known as an impulse.

4.11 Q: A tow-truck applies a force of 2000N on a 2000-kg car for a period of 3 seconds.

(A) What is the magnitude of the change in the car's momentum?

(B) If the car starts at rest, what will be its speed after 3 seconds?

4.11 A: (A) $\Delta p = F\Delta t = (2000N)(3s) = 6000N \bullet s$

(B) $\Delta p = p - p_0 = mv - mv_0$

$$v = \frac{\Delta p + mv_0}{m} = \frac{6000N \bullet s + 0}{2000kg} = 3\,{}^m\!/_s$$

4.12 Q: A 2-kilogram body is initially traveling at a velocity of 40 meters per second east. If a constant force of 10 newtons due east is applied to the body for 5 seconds, the final speed of the body is

(1) 15 m/s
(2) 25 m/s
(3) 65 m/s
(4) 130 m/s

4.12 A: (3) $Ft = \Delta p = m\Delta v$

$$\Delta v = v - v_0 = \frac{Ft}{m}$$

$$v = \frac{Ft}{m} + v_0$$

$$v = \frac{(10N)(5s)}{2kg} + 40\,{}^m\!/_s = 65\,{}^m\!/_s$$

4.13 Q: A motorcycle being driven on a dirt path hits a rock. Its 60-kilogram cyclist is projected over the handlebars at 20 meters per second into a haystack. If the cyclist is brought to rest in 0.50 seconds, the magnitude of the average force exerted on the cyclist by the haystack is

(1) 6.0×10^1 N

(2) 5.9×10^2 N

(3) 1.2×10^3 N

(4) 2.4×10^3 N

4.13 A: (4) $Ft = \Delta p = m\Delta v = m(v - v_0)$

$$F = \frac{m(v - v_0)}{t} = \frac{(60kg)(0 - 20\,{}^m\!/_s)}{0.5s} = -2400N$$

4.14 Q: The instant before a batter hits a 0.14-kilogram baseball, the velocity of the ball is 45 meters per second west. The instant after the batter hits the ball, the ball's velocity is 35 meters per second east. The bat and ball are in contact for 1.0×10^{-2} second. Calculate the magnitude of the average force the bat exerts on the ball while they are in contact.

4.14 A: $Ft = \Delta p = m\Delta v$

$$F = \frac{m\Delta v}{t} = \frac{m(v - v_0)}{t}$$

$$F = \frac{(0.14kg)(35\,{}^m\!/_s - -45\,{}^m\!/_s)}{1 \times 10^{-2}s} = 1120N$$

4.15 Q: In an automobile collision, a 44-kilogram passenger moving at 15 meters per second is brought to rest by an air bag during a 0.10-second time interval. What is the magnitude of the average force exerted on the passenger during this time?

(1) 440 N

(2) 660 N

(3) 4400 N

(4) 6600 N

4.15 A: (4) $Ft = \Delta p$

$$F = \frac{\Delta p}{t} = \frac{p - p_0}{t}$$

$$F = \frac{0 - (44kg)(15\,{}^m\!/_s)}{0.1s} = -6600N$$

Non-Constant Forces

But not all forces are constant. What do you do if a changing force is applied for a period of time? In that case, you can make a graph of the force applied on the y-axis vs. time on the x-axis. The area under the Force-Time curve is the impulse, or change in momentum.

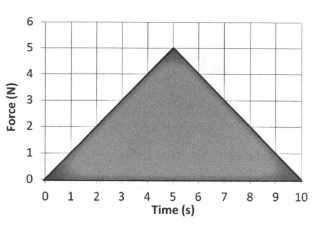

For the case of the sample graph at the right, you could determine the impulse applied by calculating the area of the triangle under the curve. In this case:

$$J = Area_{triangle} = \tfrac{1}{2}bh$$
$$J = \tfrac{1}{2}(10s)(5N) = 25N \bullet s$$

Conservation of Momentum

In an isolated system, where no external forces act, momentum is always conserved. Put more simply, in any closed system, the total momentum of the system remains constant.

In the case of a collision or explosion (an event), if you add up the individual momentum vectors of all of the objects before the event, you'll find that they are equal to the sum of the momentum vectors of the objects after the event. Written mathematically, the law of conservation of momentum states:

$$p_{initial} = p_{final}$$

This is a direct outcome of Newton's 3rd Law.

In analyzing collisions and explosions, a momentum table can be a powerful tool for problem solving. To create a momentum table, follow these basic steps:

1. Identify all objects in the system. List them vertically down the left-hand column.
2. Determine the momenta of the objects before the event. Use

variables for any unknowns.
3. Determine the momenta of the objects after the event. Use variables for any unknowns.
4. Add up all the momenta from before the event and set them equal to the momenta after the event.
5. Solve your resulting equation for any unknowns.

A **collision** is an event in which two or more objects approach and interact strongly for a brief period of time. Let's look at how the problem-solving strategy can be applied to a simple collision:

4.16 Q: A 2000-kg car traveling at 20 m/s collides with a 1000-kg car at rest at a stop sign. If the 2000-kg car has a velocity of 6.67 m/s after the collision, find the velocity of the 1000-kg car after the collision.

4.16 A: Call the 2000-kg car Car A, and the 1000-kg car Car B. You can then create a momentum table as shown below:

Objects	Momentum Before (kg·m/s)	Momentum After (kg·m/s)
Car A	2000×20=40,000	2000×6.67=13,340
Car B	1000×0=0	1000×v_B=1000v_B
Total	40,000	13,340+1000v_B

Because momentum is conserved in any closed system, the total momentum before the event must be equal to the total momentum after the event.

$$40,000 = 13,340 + 1000v_B$$

$$v_B = \frac{40,000 - 13,340}{1000} = 26.7 \,{}^m\!/_s$$

Not all problems are quite so simple, but problem solving steps remain consistent.

4.17 Q: On a snow-covered road, a car with a mass of 1.1×10^3 kilograms collides head-on with a van having a mass of 2.5×10^3 kilograms traveling at 8 meters per second. As a result of the collision, the vehicles lock together and immediately come to rest. Calculate the speed of the car immediately before the collision. [Neglect friction.]

4.17A: Define the car's initial velocity as positive and the van's initial velocity as negative. After the collision, the two objects become one, therefore you can combine them in the momentum table.

Objects	Momentum Before (kg·m/s)	Momentum After (kg·m/s)
Car	$1100 \times v_{car} = 1100 v_{car}$	0
Van	$2500 \times -8 = -20{,}000$	
Total	$-20{,}000 + 1100 v_{car}$	0

$$-20000 + 1100 v_{car} = 0$$

$$v_{car} = \frac{20000}{1100} = 18.2 \, m/s$$

4.18 Q: A 70-kilogram hockey player skating east on an ice rink is hit by a 0.1-kilogram hockey puck moving toward the west. The puck exerts a 50-newton force toward the west on the player. Determine the magnitude of the force that the player exerts on the puck during this collision.

4.18 A: The player exerts a force of 50 newtons toward the east on the puck due to Newton's 3rd Law.

4.19 Q: The diagram below represents two masses before and after they collide. Before the collision, mass m_A is moving to the right with speed v, and mass m_B is at rest. Upon collision, the two masses stick together.

Before Collision

After Collision

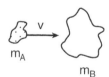

(1) $\dfrac{m_A + m_B \, v}{m_A}$

(3) $\dfrac{m_B \, v}{m_A + m_B}$

(2) $\dfrac{m_A + m_B}{m_A \, v}$

(4) $\dfrac{m_A \, v}{m_A + m_B}$

Which expression represents the speed, v′, of the masses after the collision? [Assume no outside forces are acting on m_A or m_B.]

4.19 A: Use the momentum table to set up an equation utilizing conservation of momentum, then solve for the final velocity of the combined mass, labeled v′.

Objects	Momentum Before (kg·m/s)	Momentum After (kg·m/s)
Mass A	$m_A v$	$(m_A+m_B)v'$
Mass B	0	
Total	$m_A v$	$(m_A+m_B)v'$

(4) $m_A v = (m_A + m_B)v'$

$$v' = \frac{m_A v}{m_A + m_B}$$

Let's take a look at another example which emphasizes the vector nature of momentum while examining an explosion. In physics terms, an **explosion** results when an object is broken up into two or more fragments.

4.20 Q: A 4-kilogram rifle fires a 20-gram bullet with a velocity of 300 m/s. Find the recoil velocity of the rifle.

4.20 A: Once again, you can use a momentum table to organize your problem-solving. To fill out the table, you must recognize that the initial momentum of the system is 0, and you can consider the rifle and bullet as a single system with a mass of 4.02 kg:

Objects	Momentum Before (kg·m/s)	Momentum After (kg·m/s)
Rifle	0	$4\times v_{recoil}$
Bullet		$(.020)(300)=6$
Total	0	$6+4\times v_{recoil}$

Due to conservation of momentum, you can again state that the total momentum before must equal the total momentum after, or $0=4v_{recoil}+6$. Solving for the recoil velocity of the rifle, you find:

$$0 = 4v_{recoil} + 6$$

$$v_{recoil} = \frac{-6}{4} = -1.5\,^m\!/_s$$

The negative recoil velocity indicates the direction of the rifle's velocity. If the bullet traveled forward at 300 m/s, the rifle must travel in the opposite direction.

4.21 Q: The diagram below shows two carts that were initially at rest on a horizontal, frictionless surface being pushed apart when a compressed spring attached to one of the carts is released. Cart A has a mass of 3.0 kilograms and cart B has a mass of 5.0 kilograms.

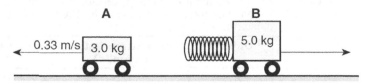

If the speed of cart A is 0.33 meter per second after the spring is released, what is the approximate speed of cart B after the spring is released?

(1) 0.12 m/s

(2) 0.20 m/s

(3) 0.33 m/s

(4) 0.55 m/s

4.21 A: Define the positive direction toward the right of the page.

Objects	Momentum Before (kg·m/s)	Momentum After (kg·m/s)
Cart A	0	3×-0.33=-1
Cart B	0	5×v$_B$
Total	0	5v$_B$-1

(2) $0 = 5v_B - 1$

$$v_B = \frac{1}{5} = 0.2 \,{}^m\!/_s$$

4.22 Q: A woman with horizontal velocity v$_1$ jumps off a dock into a stationary boat. After landing in the boat, the woman and the boat move with velocity v$_2$. Compared to velocity v$_1$, velocity v$_2$ has

(1) the same magnitude and the same direction

(2) the same magnitude and opposite direction

(3) smaller magnitude and the same direction

(4) larger magnitude and the same direction

4.22 A: (3) smaller magnitude and the same direction due to the law of conservation of momentum.

Types of Collisions

When objects collide, a number of different things can happen depending on the characteristics of the colliding objects. Of course, you know that momentum is always conserved in a closed system. Imagine, though, the differences in a collision if the two objects colliding are super-bouncy balls compared to two lumps of clay. In the first case, the balls would bounce off each other. In the second, they would stick together and become, in essence, one object. Obviously, you need more ways to characterize collisions.

Elastic collisions occur when the colliding objects bounce off of each other. This typically occurs when you have colliding objects which are very hard or bouncy. Officially, an elastic collision is one in which the sum of the kinetic energy of all the colliding objects before the event is equal to the sum of the kinetic energy of all the objects after the event. Put more simply, kinetic energy is conserved in an elastic collisions.

NOTE: There is no law of conservation of kinetic energy -- IF kinetic energy is conserved in a collision, it is called an elastic collision, but there is no physical law that requires this.

Inelastic collisions occur when two objects collide and kinetic energy is not conserved. In this type of collision some of the initial kinetic energy is converted into other types of energy (heat, sound, etc.), which is why kinetic energy is NOT conserved in an inelastic collision. In a perfectly inelastic collision, the two objects colliding stick together.

In reality, most collisions fall somewhere between the extremes of a completely elastic collision and a completely inelastic collision.

4.23 Q: Two billiard balls collide. Ball 1 moves with a velocity of 4 m/s, and ball 2 is at rest. After the collision, ball 1 comes to a complete stop. What is the velocity of ball 2 after the collision? Is this collision elastic or inelastic? The mass of each ball is 0.16 kg.

4.23 A: To find the velocity of ball 2, use a momentum table.

Objects	Momentum Before (kg·m/s)	Momentum After (kg·m/s)
Ball 1	0.16×4=0.64	0
Ball 2	0	$0.16 \times v_2$
Total	0.64	$0.16 \times v_2$

$$0.64 = 0.16 \times v_2$$

$$v_2 = \frac{0.64 \frac{kg \times m}{s}}{0.16 kg} = 4 \, m\!/\!_s$$

To determine whether this is an elastic or inelastic collision, you can calculate the total kinetic energy of the system both before and after the collision.

Objects	KE Before (J)	KE After (J)
Ball 1	0.5*0.16×4²=1.28	0
Ball 2	0	0.5*0.16×4²=1.28
Total	1.28	1.28

Since the kinetic energy before the collision is equal to the kinetic energy after the collision (kinetic energy is conserved), this is an elastic collision.

4.24 Q: A baseball and a bowling ball are rolling along a flat surface with equal momenta. How do the velocities of the balls compare?

(1) The baseball has a higher velocity than the bowling ball.

(2) The bowling ball has a higher velocity than the baseball.

(3) They are the same.

4.24 A: (1) Because they have the same momenta, and the baseball has a smaller mass, the baseball must have a higher velocity.

4.25 Q: A baseball and a bowling ball are rolling along a flat surface with equal momenta. An equal force is exerted on each to stop their motion. Which ball takes longer to come to a complete stop?

(1) The baseball takes longer to stop.

(2) The bowling ball takes longer to stop.

(3) They take the same amount of time to stop.

4.25 A: (3) They take the same amount of time to stop since an equal force is applied, and they start with the same momenta.

$$J = \Delta p = F \Delta t \rightarrow \Delta t = \frac{\Delta p}{F}$$

Since the change in momentum is the same, and the force is the same, the time the force is applied must also be the same.

Collisions in 2 Dimensions

Much like the key to projectile motion, or two-dimensional kinematics problems, was breaking up vectors into their x- and y-components, the key to solving two-dimensional collision problems involves breaking up momentum vectors into x- and y- components. The law of conservation of momentum then states that momentum is independently conserved in both the x- and y- directions.

$$p_{initial_x} = p_{final_x}$$

$$p_{initial_y} = p_{final_y}$$

Therefore, you can solve two-dimensional collision problems by creating a separate momentum table for the x-component of momentum before and after the collision, and a momentum table for the y-component of momentum.

4.26 Q: Bert strikes a cue ball of mass 0.17 kg, giving it a velocity of 3 m/s in the x-direction. When the cue ball strikes the eight ball (mass=0.16 kg), previously at rest, the eight ball is deflected 45 degrees from the cue ball's previous path, and the cue ball is deflected 40 degrees in the opposite direction. Find the velocity of the cue ball and the eight ball after the collision.

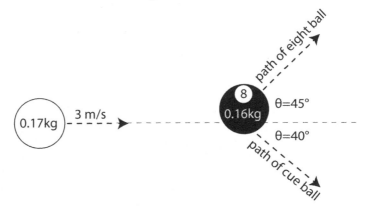

4.26 A: Start by making momentum tables for the collision, beginning with the x-direction. Since you don't know the velocity of the balls after the collision, call the velocity of the cue ball after the collision v_c, and the velocity of the eight ball after the collision v_8. Note that you must use trigonometry to determine the x-component of the momentum of each ball after the collision.

Objects	X-Momentum Before (kg·m/s)	X-Momentum After (kg·m/s)
Cue Ball	0.17×3=0.51	0.17×v_c×cos(40°)
Eight Ball	0	0.16×v_8×cos(45°)
Total	0.51	0.17×v_c×cos(40°)+ 0.16×v_8×cos(45°)

Since the total momentum in the x-direction before the collision must equal the total momentum in the x-direction after the collision, you can set the total before and total after columns equal:

$$0.51^{kg\bullet m}\!/_{s} = (0.17kg)(\cos 40°)v_c + (0.16kg)(\cos 45°)v_8$$

$$0.51^{kg\bullet m}\!/_{s} = (0.130kg)v_c + (0.113kg)v_8$$

Next, create a momentum table and algebraic equation for the conservation of momentum in the y-direction.

Objects	Y-Momentum Before (kg·m/s)	Y-Momentum After (kg·m/s)
Cue Ball	0	0.17×v_c×sin(-40°)
Eight Ball	0	0.16×v_8×sin(45°)
Total	0	0.17×v_c×sin(-40°)+ 0.16×v_8×sin(45°)

$$0 = (0.17kg)(\sin- 40°)v_c + (0.16kg)(\sin 45°)v_8$$

$$0 = (-0.109kg)v_c + (0.113kg)v_8$$

You now have two equations with two unknowns. To solve this system of equations, start by solving the y-momentum equation for v_c.

$$0 = (-0.109kg)v_c + (0.113kg)v_8$$

$$(0.109kg)v_c = (0.113kg)v_8$$

$$v_c = 1.04v_8$$

You can now take this equation for v_c and substitute it into the equation for conservation of momentum in the x-direction, effectively eliminating one of the unknowns, and giving a single equation with a single unknown.

$$0.51^{kg\bullet m}\!/_{s} = (0.130kg)v_c + (0.113kg)v_8 \xrightarrow{v_c=1.04v_8}$$

$$0.51^{kg\bullet m}\!/_{s} = (0.130kg)(1.04v_8)+(0.113kg)v_8$$

$$0.51^{kg\bullet m}\!/_{s} = (0.248kg)v_8$$

$$v_8 = 2.06^{m}\!/_{s}$$

Finally, solve for the velocity of the cue ball after the collision by substituting the known value for v_8 into the result of the y-momentum equation.

$$v_c = 1.04v_8 \xrightarrow{v_8=2.06^{m}\!/_{s}}$$

$$v_c = (1.04)(2.06^{m}\!/_{s}) = 2.14^{m}\!/_{s}$$

Center of Mass

The motion of real objects is considerably more complex than that of simple theoretical particles. However, you can treat an entire object as if its entire mass were contained at a single point, known as the object's **center of mass** (CM). Mathematically, the center of mass of an object is the weighted average of the location of mass in an object.

You can find the center of mass of a system of particles by taking the sum of the mass of the particles, multiplied by their positions, and dividing that by the total mass of the object. Looking at this in two dimensions, the center of mass in the x- and y-directions would be:

$$x_{CM} = \frac{m_1 x_1 + m_2 x_2 + ...}{m_1 + m_2 + ...}$$

$$y_{CM} = \frac{m_1 y_1 + m_2 y_2 + ...}{m_1 + m_2 + ...}$$

No matter how complex an object may be, you can calculate its center of mass and then treat the object as a point particle with total mass M. This allows you to apply basic physics principles to complex objects without adding unnecessary mathematical complexity to the analysis!

4.27 Q: Find the center of mass of an object modeled as two separate masses on the x-axis. The first mass is 2 kg at an x-coordinate of 2 and the second mass is 6 kg at an x-coordinate of 8.

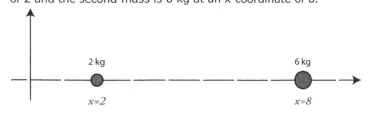

4.27 A:
$$x_{CM} = \frac{m_1 x_1 + m_2 x_2 + ...}{m_1 + m_2 + ...}$$

$$x_{CM} = \frac{(2kg)(2) + (6kg)(8)}{(2kg + 6kg)} = 6.5$$

This means you can treat the object as a point particle with a mass of 8 kg at an x-coordinate of 6.5 as shown below.

Use the same strategy for finding the center of mass of a multi-dimensional object.

4.28 Q: Find the coordinates of the center of mass for the system shown below.

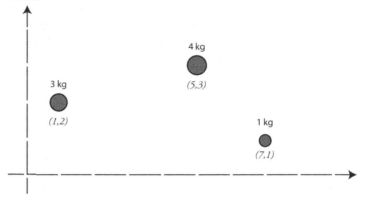

4.28 A:
$$x_{CM} = \frac{m_1x_1 + m_2x_2 + \ldots}{m_1 + m_2 + \ldots} = \frac{(3kg)(1) + (4kg)(5) + (1kg)(7)}{(3kg + 4kg + 1kg)} = 3.75$$

$$y_{CM} = \frac{m_1y_1 + m_2y_2 + \ldots}{m_1 + m_2 + \ldots} = \frac{(3kg)(2) + (4kg)(3) + (1kg)(1)}{(3kg + 4kg + 1kg)} = 2.38$$

Therefore the center of mass is a point particle with mass 8 kg located at (3.75, 2.38).

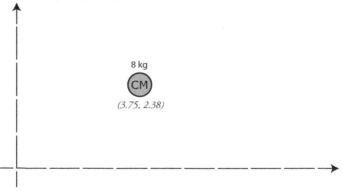

Note that center of mass is not the same as center of gravity. In a uniform gravitational field, they are the same, but center of gravity refers to the location at which the force of gravity acts upon an object as if it were a point particle with all its mass focused at that point, a subtle but important difference.

Chapter 5: Uniform Circular Motion & Gravity

"I can calculate the motion of heavenly bodies, but not the madness of people."

— Sir Isaac Newton

Objectives

1. Explain the acceleration of an object moving in a circle at constant speed.
2. Define centripetal force and recognize that it is not a special kind of force, but that it is provided by forces such as tension, gravity, and friction.
3. Solve problems involving calculations of centripetal force.
4. Determine the direction of a centripetal force and centripetal acceleration for an object moving in a circular path.
5. Calculate the period, frequency, speed and distance traveled for objects moving in circles at constant speed.
6. Analyze and solve problems involving objects moving in vertical circles.
7. Determine the acceleration due to gravity near the surface of Earth.
8. Utilize Newton's Law of Universal Gravitation to determine the gravitational force of attraction between two objects.
9. Explain the difference between mass and weight.
10. Explain weightlessness for objects in orbit.
11. Explain how Kepler's Laws describe the orbits of planetary objects around the sun.

Now that you've talked about linear and projectile kinematics, as well as fundamentals of dynamics and Newton's Laws, you have the skills and background to analyze circular motion. Of course, this has obvious applications such as cars moving around a circular track, roller coasters moving in loops, and toy trains chugging around a circular track under the Christmas tree. Less obvious, however, is the application to turning objects. Any object that is turning can be thought of as moving through a portion of a circular path, even if it's just a small portion of that path.

With this realization, analysis of circular motion will allow you to explore a car speeding around a corner on an icy road, a running back cutting upfield to avoid a blitzing linebacker, and the orbits of planetary bodies. The key to understanding all of these phenomena starts with the study of uniform circular motion.

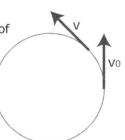

Centripetal Acceleration

The motion of an object in a circular path at constant speed is known as **uniform circular motion** (**UCM**). An object in UCM is constantly changing direction, and since velocity is a vector and has direction, you could say that an object undergoing UCM has a constantly changing velocity, even if its speed remains constant. And if the velocity of an object is changing, it must be accelerating. Therefore, an object undergoing UCM is constantly accelerating. This type of acceleration is known as **centripetal acceleration**.

5.01 Q: If a car is accelerating, is its speed increasing?

5.01 A: It depends. Its speed could be increasing, or it could be accelerating in a direction opposite its velocity (slowing down). Or, its speed could remain constant yet still be accelerating if it is traveling in uniform circular motion.

Just as importantly, you'll need to figure out the direction of the object's acceleration, since acceleration is a vector. To do this, draw an object moving counter-clockwise in a circular path, and show its velocity vector at two different points in time. Since acceleration is the rate of change of an object's velocity with respect to time, you can determine the direction of the object's acceleration by finding the direction of its change in velocity, Δv.

To find its change in velocity, Δv, recall that $\Delta v = v - v_0$.

Therefore, you can find the difference of the vectors v and v_0 graphically, which can be re-written as $\Delta v = v + (-v_0)$.

Recall that to add vectors graphically, you line them up tip-to-tail, then draw the resultant vector from the starting point (tail) of the first vector to the ending point (tip) of the last vector.

So, the acceleration vector must point in the direction shown above. If this vector is shown back on the original circle, lined up directly between the initial and final velocity vector, it's easy to see that the acceleration vector points toward the center of the circle.

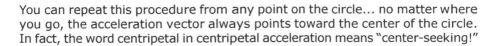

You can repeat this procedure from any point on the circle… no matter where you go, the acceleration vector always points toward the center of the circle. In fact, the word centripetal in centripetal acceleration means "center-seeking!"

So now that you know the direction of an object's acceleration (toward the center of the circle), what about its magnitude? The formula for the magnitude of an object's centripetal acceleration is given by:

$$a_c = \frac{v^2}{r}$$

5.02 Q: In the diagram below, a cart travels clockwise at constant speed in a horizontal circle.

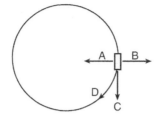

At the position shown in the diagram, which arrow indicates the direction of the centripetal acceleration of the cart?

(1) A

(2) B

(3) C

(4) D

5.02 A: (1) The acceleration of any object moving in a circlular path is toward the center of the circle.

5.03 Q: The diagram shows the top view of a 65-kilogram student at point A on an amusement park ride. The ride spins the student in a horizontal circle of radius 2.5 meters, at a constant speed of 8.6 meters per second. The floor is lowered and the student remains against the wall without falling to the floor.

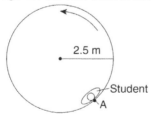

Which vector best represents the direction of the centripetal acceleration of the student at point A?

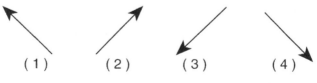

5.03 A: (1) Centripetal acceleration points toward the center of the circle.

Chapter 5: Uniform Circular Motion & Gravity

5.04 Q: Which graph best represents the relationship between the magnitude of the centripetal acceleration and the speed of an object moving in a circle of constant radius?

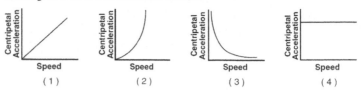

(1)   (2)   (3)   (4)

5.04 A: (2) Centripetal acceleration is proportional to v²/r.

5.05 Q: A car rounds a horizontal curve of constant radius at a constant speed. Which diagram best represents the directions of both the car's velocity, v, and acceleration, a?

5.05 A: (3) Velocity is tangent to the circular path, and acceleration is toward the center of the circular path.

5.06 Q: A 0.50-kilogram object moves in a horizontal circular path with a radius of 0.25 meter at a constant speed of 4.0 meters per second. What is the magnitude of the object's acceleration?

(1) 8 m/s²

(2) 16 m/s²

(3) 32 m/s²

(4) 64 m/s²

5.06 A: (4) 64 m/s².

Circular Speed

So how do you find the speed of an object as it travels in a circular path? The formula for speed that you learned in kinematics still applies.

$$\overline{v} = \frac{x}{t}$$

You have to be careful in using this equation, however, to understand that an object traveling in a circular path is traveling along the circumference of a circle. Therefore, if an object were to make one complete revolution around the circle, the distance it travels is equal to the circle's circumference.

$$C = 2\pi r$$

5.07 Q: Miranda drives her car clockwise around a circular track of radius 30m. She completes 10 laps around the track in 2 minutes. Find Miranda's total distance traveled, average speed, and centripetal acceleration.

5.07 A: $d = 10 \times 2\pi r = 10 \times 2\pi(30m) = 1885m$

$$\overline{v} = \frac{x}{t} = \frac{1885m}{120s} = 15.7\,^m\!/_s$$

$$a_c = \frac{v^2}{r} = \frac{(15.7\,^m\!/_s)^2}{30m} = 8.2\,^m\!/_{s^2}$$

5.08 Q: The combined mass of a race car and its driver is 600 kilograms. Traveling at constant speed, the car completes one lap around a circular track of radius 160 meters in 36 seconds. Calculate the speed of the car.

5.08 A: $\overline{v} = \dfrac{x}{t} = \dfrac{2\pi r}{t} = \dfrac{2\pi(160m)}{36s} = 27.9\,^m\!/_s$

Centripetal Force

If an object traveling in a circular path has an inward acceleration, Newton's 2nd Law states there must be a net force directed toward the center of the circle as well. This type of force, known as a **centripetal force**, can be a gravitational force, a tension, an applied force, or even a frictional force.

NOTE: When dealing with circular motion problems, it is important to realize that a centripetal force isn't really a new force, a centripetal force is just a label or grouping you apply to a force to indicate its direction is toward the center of a circle. This means that you never want to label a force on a free body diagram as a centripetal force, F_c. Instead, label the center-directed force as specifically as you can. If a tension is causing the force, label the force F_T. If a frictional force is causing the center-directed force, label it F_f, and so forth.

You can combine the equation for centripetal acceleration with Newton's 2nd Law to obtain Newton's 2nd Law for Circular Motion. Recall that Newton's 2nd Law states:

$$F_{net} = ma$$

For an object traveling in a circular path, there must be a net (centripetal) force directed toward the center of the circular path to cause a (centripetal) acceleration directed toward the center of the circular path. You can revise Newton's 2nd Law for this particular case as follows:

$$F_C = ma_C$$

Then, recalling the formula for centripetal acceleration as:

$$a_c = \frac{v^2}{r}$$

You can put these together, replacing a_c in the equation to get a combined form of Newton's 2nd Law for Uniform Circular Motion:

$$F_C = \frac{mv^2}{r}$$

Of course, if an object is traveling in a circular path and the centripetal force is removed, the object will continue traveling in a straight line in whatever direction it was moving at the instant the force was removed.

5.09 Q: An 800N running back turns a corner in a circular path of radius 1 meter at a velocity of 8 m/s. Find the running back's mass, centripetal acceleration, and centripetal force.

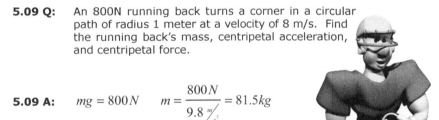

5.09 A: $mg = 800N \qquad m = \dfrac{800N}{9.8\,{}^{m}\!/\!_{s^2}} = 81.5kg$

$a_c = \dfrac{v^2}{r} = \dfrac{(8\,{}^{m}\!/\!_{s})^2}{1m} = 64\,{}^{m}\!/\!_{s^2}$

$F_c = ma_c = (81.5kg)(64\,{}^{m}\!/\!_{s^2}) = 5220N$

5.10 Q: The diagram at right shows a 5.0-kilogram bucket of water being swung in a horizontal circle of 0.70-meter radius at a constant speed of 2.0 meters per second. The magnitude of the centripetal force on the bucket of water is approximately

(1) 5.7 N

(2) 14 N

(3) 29 N

(4) 200 N

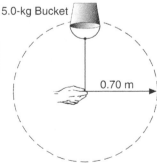

5.0-kg Bucket

0.70 m

5.10 A: (3) $F_c = ma_c = m\dfrac{v^2}{r} = (5kg)\dfrac{(2\,^m\!/_s)^2}{0.7m} = 29N$

5.11 Q: A 1.0 × 10³-kilogram car travels at a constant speed of 20 meters per second around a horizontal circular track. Which diagram correctly represents the direction of the car's velocity (v) and the direction of the centripetal force (F_c) acting on the car at one particular moment?

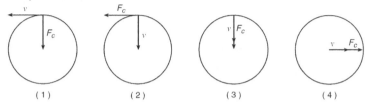

       (1)   (2)   (3)   (4)

5.11 A: (1) Velocity is tangent to the circle, and the centripetal force points toward the center of the circle.

5.12 Q: A 1750-kilogram car travels at a constant speed of 15 meters per second around a horizontal, circular track with a radius of 45 meters. The magnitude of the centripetal force acting on the car is

(1) 5.00 N

(2) 583 N

(3) 8750 N

(4) 3.94 10⁵ N

5.12 A: (3) $F_c = ma_c = m\dfrac{v^2}{r} = (1750kg)\dfrac{(15\,^m\!/_s)^2}{45m} = 8750N$

5.13 Q: A ball attached to a string is moved at constant speed in a horizontal circular path. A target is located near the path of the ball as shown in the diagram. At which point along the ball's path should the string be released, if the ball is to hit the target?

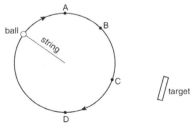

(1) A

(2) B

(3) C

(4) D

5.13 A: (B) If released at point B, the ball will continue in a straight line to the target.

5.14 Q: A 1200-kilogram car traveling at a constant speed of 9 meters per second turns at an intersection. The car follows a horizontal circular path with a radius of 25 meters to point P. At point P, the car hits an area of ice and loses all frictional force on its tires. Which path does the car follow on the ice?

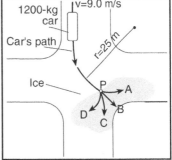

(1) A

(2) B

(3) C

(4) D

5.14 A: (2) Once the car loses all frictional force, there is no longer a force toward the center of the circular path, therefore the car will travel in a straight line toward B.

Frequency and Period

For objects moving in circular paths, you can characterize their motion around the circle using the terms frequency (f) and period (T). The **frequency** of an object is the number of revolutions the object makes in a complete second. It is measured in units of [1/s], or Hertz (Hz). In similar fashion, the **period** of an object is the time it takes to make one complete revolution. Since the period is a time interval, it is measured in units of seconds. You can relate period and frequency using the equations:

$$f = \frac{1}{T} \qquad T = \frac{1}{f}$$

5.15 Q: A 500g toy train completes 10 laps of its circular track in 1 minute and 40 seconds. If the diameter of the track is 1 meter, find the train's centripetal acceleration (a_c), centripetal force (F_c), period (T), and frequency (f).

5.15 A: $\overline{v} = \dfrac{x}{t} = \dfrac{2\pi r \times 10}{t} = \dfrac{2\pi(0.5m) \times 10}{100s} = 0.314\,{}^{m}\!/_{s}$

$a_c = \dfrac{v^2}{r} = \dfrac{(0.314\,{}^{m}\!/_{s})^2}{0.5m} = 0.197\,{}^{m}\!/_{s^2}$

$F_c = ma_c = (0.5kg)(0.197\,{}^{m}\!/_{s^2}) = 0.099\,N$

$T = \dfrac{100s}{10\,revs} = 10s$

$f = \dfrac{1}{T} = \dfrac{1}{10s} = 0.1\,Hz$

5.16 Q: Alan makes 38 complete revolutions on the playground Round-A-Bout in 30 seconds. If the radius of the Round-A-Bout is 1 meter, determine

(A) Period of the motion

(B) Frequency of the motion

(C) Speed at which Alan revolves

(D) Centripetal force on 40-kg Alan

5.16 A: (A) $T = \dfrac{30s}{38\,revs} = 0.789s$

(B) $f = \dfrac{1}{T} = \dfrac{1}{0.789s} = 1.27\,Hz$

(C) $\overline{v} = \dfrac{x}{t} = \dfrac{38 \times 2\pi r}{t} = \dfrac{38 \times 2\pi(1m)}{30s} = 7.96\,{}^{m}\!/_{s}$

(D) $F_c = ma_c = m\dfrac{v^2}{r} = (40kg)\dfrac{(7.96\,{}^{m}\!/_{s})^2}{1m} = 2530\,N$

Vertical Circular Motion

Objects travel in circles vertically as well as horizontally. Because the speed of these objects isn't typically constant, technically this isn't uniform circular motion, but your UCM analysis skills still prove applicable.

Consider a roller coaster traveling in a vertical loop of radius 10m. You travel through the loop upside down, yet you don't fall out of the roller coaster. How is this possible? You can use your understanding of UCM and dynamics to find out!

To begin with, first take a look at the coaster when the car is at the bottom of the loop. Drawing a free body diagram, the force of gravity on the coaster, also known as its weight, pulls it down, so draw a vector pointing down labeled "mg." Opposing that force is the normal force of the rails of the coaster pushing up, which is labeled F_N.

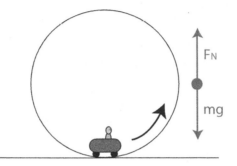

Because the coaster is moving in a circular path, you can analyze it using the tools developed for uniform circular motion. Newton's 2nd Law still applies, so you can write:

$$F_{NET_c} = F_N - mg$$

Notice that because you're talking about circular motion, you can adopt the convention that forces pointing toward the center of the circle are positive, and forces pointing away from the center of the circle are negative. At this point, recall that the force you "feel" when you're in motion is actually the normal force. So, solving for the normal force as you begin to move in a circle, you find:

$$F_N = F_{NET_c} + mg$$

Since you know that the net force is always equal to mass times acceleration, the net centripetal force is equal to mass times the centripetal acceleration. You can therefore replace F_{NETc} as follows:

$$F_N = F_{NET_c} + mg = \frac{mv^2}{r} + mg$$

You can see from the resulting equation that the normal force is now equal to the weight plus an additional term from the centripetal force of the circular motion. As you travel in a circular path near the bottom of the loop, you feel heavier than your weight. In common terms, you feel additional "g-forces." How many g's you feel can be obtained with a little bit more manipulation. If you re-write your equation for the normal force, pulling out the mass by applying the distributive property of multiplication, you obtain:

$$F_N = m\left(\frac{v^2}{r} + g\right)$$

Notice that inside the parenthesis you have the standard acceleration due to gravity, g, plus a term from the centripetal acceleration. This additional term is the additional g-force felt by a person. For example, if a_c was equal to g (9.8 m/s²), you could say the person in the cart was experiencing two g's (1g from the centripetal acceleration, and 1g from the Earth's gravitational field). If a_c were equal to 3×g (29.4 m/s²), the person would be experiencing a total of four g's.

Expanding this analysis to a similar situation in a different context, try to imagine instead of a roller coaster, a mass whirling in a vertical circle by a string. You could replace the normal force by the tension in the string in the analysis. Because the force is larger at the bottom of the circle, the likelihood of the string breaking is highest when the mass is at the bottom of the circle!

At the top of the loop, you have a considerably different picture. Now, the normal force from the coaster rails must be pushing down against the cart, though still in the positive direction since down is now toward the center of the circular path. In this case, however, the weight of the object also points toward the center of the circle, since the Earth's gravitational field always pulls toward the center of the Earth. The free body diagram looks considerably different, and therefore the application of Newton's 2nd Law for Circular Motion is considerably different as well:

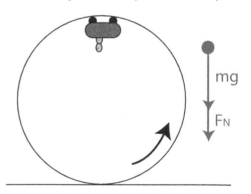

$$F_{NET_c} = F_N + mg$$

Since the force you feel is actually the normal force, you can solve for the normal force and expand the net centripetal force as shown:

$$F_N = F_{NET_c} - mg = \frac{mv^2}{r} - mg$$

You can see from the equation that the normal force is now the centripetal force minus your weight. If the centripetal force were equal to your weight, you would feel as though you were weightless. Note that this is also the point where the normal force is exactly equal to 0. This means the rails of the track are no longer pushing on the roller coaster cart. If the centripetal force was slightly smaller, and the cart's speed was slightly smaller, the normal force F_N would be less than 0. Since the rails can't physically pull the cart in the negative direction (away from the center of the circle), this means the car is falling off the rail and the cart's occupant is about to have a very, very bad day. Only by maintaining a high speed can the cart successfully negotiate the loop. Go too slow and the cart falls.

In order to remain safe, real roller coasters actually have wheels on both sides of the rails to prevent the cart from falling if it ever did slow down at the top of a loop, although coasters are designed so that this situation never actually occurs.

5.17 Q: In an experiment, a rubber stopper is attached to one end of a string that is passed through a plastic tube before weights are attached to the other end. The stopper is whirled in a horizontal circular path at constant speed.

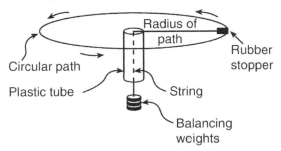

(A) Describe what would happen to the radius of the circle if the student whirls the stopper at a greater speed without changing the balancing weights.

(B) The rubber stopper is now whirled in a vertical circle at the same speed. On the vertical diagram, draw and label vectors to indicate the direction of the weight (F_g) and the direction of the centripetal force (F_c) at the position shown.

5.17 A: (A) As the speed of the stopper is increased, the radius of the orbit will increase.

(B)

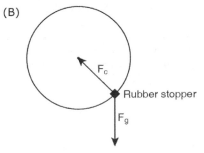

Universal Gravitation

All objects that have mass attract each other with a gravitational force. The magnitude of that force, F_g, can be calculated using Newton's Law of Universal Gravitation:

$$F_g = \frac{Gm_1 m_2}{r^2}$$

This law says that the force of gravity between two objects is proportional to each of the masses(m_1 and m_2) and inversely proportional to the square of the distance between them (r). The **universal gravitational constant**, G, is a "fudge factor," so to speak, included in the equation so that your answers come out in S.I. units. G is given as 6.67×10^{-11} N·m²/kg².

Let's look at this relationship in a bit more detail. Force is directly proportional to the masses of the two objects, therefore if either of the masses were doubled, the gravitational force would also double. In similar fashion, if the distance between the two objects, r, was doubled, the force of gravity would be quartered since the distance is squared in the denominator. This type of relationship is called an inverse square law, which describes many phenomena in the natural world.

NOTE: The distance between the masses, r, is actually the distance between the center of masses of the objects. For large objects, such as the gravitational attraction between the Earth and the moon, you must determine the distance from the center of the Earth to the center of the moon, not their surfaces.

Some hints for problem solving when dealing with Newton's Law of Universal Gravitation:

1. Substitute values in for variables at the very end of the problem only. The longer you can keep the formula in terms of variables, the fewer opportunities for mistakes.
2. Before using your calculator to find an answer, try to estimate the order of magnitude of the answer. Use this to check your final answer.
3. Once your calculations are complete, make sure your answer makes sense by comparing your answer to a known or similar quantity. If your answer doesn't make sense, check your work and verify your calculations.

5.18 Q: What is the gravitational force of attraction between two asteroids in space, each with a mass of 50,000 kg, separated by a distance of 3800 m?

5.18 A:

$$F_g = \frac{Gm_1m_2}{r^2}$$

$$F_g = \frac{(6.67 \times 10^{-11} \, {N \bullet m^2}/{kg^2})(50000kg)(50000kg)}{(3800m)^2} = 1.15 \times 10^{-8} \, N$$

As you can see, the force of gravity is a relatively weak force, and you would expect a relatively weak force between relatively small objects. It takes tremendous masses and relatively small distances in order to develop significant gravitational forces. Let's take a look at another problem to explore the relationship between gravitational force, mass, and distance.

5.19 Q: As a meteor moves from a distance of 16 Earth radii to a distance of 2 Earth radii from the center of Earth, the magnitude of the gravitational force between the meteor and Earth becomes

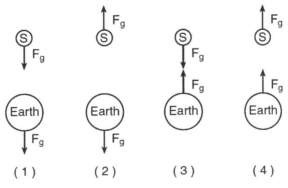

(1) one-eighth as great

(2) 8 times as great

(3) 64 times as great

(4) 4 times as great

5.19 A: (3) 64 times as great. The gravitational force is given by Newton's Law of Universal Gravitation. If the radius is one-eighth its initial value, and radius is squared in the denominator, the radius squared becomes one-sixty fourth its initial value. Because radius squared is in the denominator, the gravitational force must increase by 64X.

5.20 Q: Which diagram best represents the gravitational forces, F_g, between a satellite, S, and Earth?

5.20 A: (3) Newton's 3rd Law says that the force of gravity on the satellite due to Earth will be equal in magnitude and opposite in the direction the force of gravity on the Earth due to the satellite.

5.21 Q: Io (pronounced "EYE oh") is one of Jupiter's moons discovered by Galileo. Io is slightly larger than Earth's Moon. The mass of Io is 8.93×10^{22} kilograms and the mass of Jupiter is 1.90×10^{27} kilograms. The distance between the centers of Io and Jupiter is 4.22×10^8 meters.

(A) Calculate the magnitude of the gravitational force of attraction that Jupiter exerts on Io.

(B) Calculate the magnitude of the acceleration of Io due to the gravitational force exerted by Jupiter.

5.21 A: (A) 6.35×10^{22} N

$$F_g = \frac{Gm_1m_2}{r^2} = \frac{(6.67 \times 10^{-11}\ {}^{N \bullet m^2}\!/_{kg^2})(8.93 \times 10^{22}\ kg)(1.9 \times 10^{27}\ kg)}{(4.22 \times 10^8\ m)^2}$$

$$F_g = 6.35 \times 10^{22}\ N$$

(B) $a = \dfrac{F_{net}}{m} = \dfrac{6.35 \times 10^{22}\ N}{8.93 \times 10^{22}\ kg} = 0.71\ {}^{m}\!/_{s^2}$

5.22 Q: The diagram shows two bowling balls, A and B, each having a mass of 7 kilograms, placed 2 meters apart.

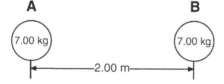

A   B

7.00 kg   7.00 kg

|←——2.00 m——→|

What is the magnitude of the gravitational force exerted by ball A on ball B?

(1) 8.17×10^{-9} N

(2) 1.63×10^{-9} N

(3) 8.17×10^{-10} N

(4) 1.17×10^{-10} N

5.22 A: (3) $F_g = \dfrac{Gm_1m_2}{r^2} = \dfrac{(6.67 \times 10^{-11}\ {}^{N \bullet m^2}\!/_{kg^2})(7 kg)(7 kg)}{(2m)^2} = 8.17 \times 10^{-10}\ N$

5.23 Q: A 2.0-kilogram object is falling freely near Earth's surface. What is the magnitude of the gravitational force that Earth exerts on the object?

(1) 20 N
(2) 2.0 N
(3) 0.20 N
(4) 0.0 N

5.23 A: (1) 20 N

Gravitational Fields

Gravity is a non-contact, or field, force. Its effects are observed without the two objects coming into contact with each other. Exactly how this happens is a mystery to this day, but scientists have come up with a mental construct to help understand how gravity works.

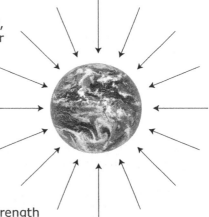

Envision an object with a gravitational field, such as the planet Earth. The closer other masses are to Earth, the more gravitational force they will experience. You can characterize this by calculating the amount of force the Earth will exert per unit mass at various distances from the Earth. Obviously, the closer the object is to the Earth, the larger a force it will experience, and the farther it is from the Earth, the smaller a force it will experience.

Attempting to visualize this, picture the strength of the gravitational force on a test object represented by a vector at the position of the object. The denser the force vectors are, the stronger the force, the stronger the "gravitational field." As these field lines become less dense, the gravitational field gets weaker.

To calculate the gravitational field strength at a given position, go back to the definition of the force of gravity on the test object, better known as its weight. You've been writing this as mg since beginning your study of dynamics. Realizing that this is the force of gravity on an object, you can also calculate the force of gravity on a test mass using Newton's Law of Universal Gravitation. Putting these together you find that:

$$F_g = mg = \frac{Gm_1m_2}{r^2}$$

Realizing that the mass on the left-hand side of the equation, the mass of the test object, is also one of the masses on the right-hand side of the equation, you can simplify the expression by dividing out the test mass.

$$g = \frac{Gm}{r^2}$$

Therefore, the gravitational field strength, g, is equal to the universal gravitational constant, G, times the mass of the object, divided by the square of the distance between the objects.

But wait, you might say, I thought g was the acceleration due to gravity on the surface of the Earth! And you would be right. Not only is g the gravitational field strength, it's also the acceleration due to gravity. The units even work out. The units of gravitational field strength, N/kg, are equivalent to the units for acceleration, m/s²!

Still skeptical? Try to calculate the gravitational field strength on the surface of the Earth using the knowledge that the mass of the Earth is approximately 5.98×10²⁴ kg and the distance from the surface to the center of mass of the Earth (which varies slightly since the Earth isn't a perfect sphere) is approximately 6378 km.

$$g = \frac{Gm}{r^2} = \frac{(6.67 \times 10^{-11} \ N \bullet m^2/_{kg^2})(5.98 \times 10^{24} \ kg)}{(6378000m)^2} = 9.8 \ m/_{s^2}$$

As expected, the gravitational field strength on the surface of the Earth is the acceleration due to gravity.

5.24 Q: Suppose a 100-kg astronaut feels a gravitational force of 700N when placed in the gravitational field of a planet.

A) What is the gravitational field strength at the location of the astronaut?

B) What is the mass of the planet if the astronaut is 2×10⁶ m from its center?

5.24 A: (A) $F_g = mg$

$$g = \frac{F_g}{m} = \frac{700N}{100kg} = 7 \ N/_{kg}$$

(B) $F_g = \frac{Gm_1 m_2}{r^2}$ $\quad m_{planet} = \frac{F_g r^2}{Gm_{astronaut}}$

$$m_{planet} = \frac{(700N)(2 \times 10^6 m)^2}{(6.67 \times 10^{-11} \ N \bullet m^2/_{kg^2})(100kg)} = 4.2 \times 10^{23} \ kg$$

5.25 Q: What is the acceleration due to gravity at a location where a 15-kilogram mass weighs 45 newtons?

(1) 675 m/s²

(2) 9.81 m/s²

(3) 3.00 m/s²

(4) 0.333 m/s²

5.25 A: (3) $F_g = mg$

$$g = \frac{F_g}{m} = \frac{45N}{15kg} = 3\,{}^{N}\!/_{kg} = 3\,{}^{m}\!/_{s^2}$$

5.26 Q: A 1200-kilogram space vehicle travels at 4.8 meters per second along the level surface of Mars. If the magnitude of the gravitational field strength on the surface of Mars is 3.7 newtons per kilogram, the magnitude of the normal force acting on the vehicle is

(1) 320 N

(2) 930 N

(3) 4400 N

(4) 5800 N

5.26 A: (3) To solve this problem, you must recognize that the gravitational field strength of 3.7 N/kg is equivalent to the acceleration due to gravity on Mars, therefore a=3.7 m/s². Then, because the space vehicle isn't accelerating vertically, the normal force must balance the vehicle's weight.

$$F_N = F_g = mg = (1200kg)(3.7\,{}^{m}\!/_{s^2}) = 4440N$$

5.27 Q: A 2.00-kilogram object weighs 19.6 newtons on Earth. If the acceleration due to gravity on Mars is 3.71 meters per second², what is the object's mass on Mars?

(1) 2.64 kg

(2) 2.00 kg

(3) 19.6 N

(4) 7.42 N

5.27 A: (2) 2.00 kg. Mass does not change.

5.28 Q: The graph below represents the relationship between gravitational force and mass for objects near the surface of Earth.

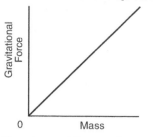

The slope of the graph represents the

(1) acceleration due to gravity

(2) universal gravitational constant

(3) momentum of objects

(4) weight of objects

5.28 A: (1) acceleration due to gravity.

Orbits

How do celestial bodies orbit each other? The moon orbits the Earth. The Earth orbits the sun. Earth's solar system is in orbit in the Milky Way galaxy... but how does it all work?

To explain orbits, Sir Isaac Newton developed a "thought experiment" in which he imagined a cannon placed on top of a very tall mountain, so tall, in fact, that the peak of the mountain was above the atmosphere (this is important because it allows us to neglect air resistance). If the cannon then launched a projectile horizontally, the projectile would follow a parabolic path to the surface of the Earth.

If the projectile was launched with a higher speed, however, it would travel farther across the surface of the Earth before reaching the ground. If its speed could be increased high enough, the projectile would fall at the same rate the Earth's surface curves away. The projectile would continue falling forever as it circled the Earth! This circular motion describes an orbit.

Put another way, the astronauts in the Space Shuttle aren't weightless. Far from it, actually, the Earth's gravity is still acting on them and pulling them toward the center of the Earth with a substantial force. You can even calculate that force.

5.29 Q: If the Space Shuttle orbits the Earth at an altitude of 380 km, what is the gravitational field strength due to the Earth?

5.29 A: Recall that values for G, the mass of the Earth, and the radius of the Earth are well-known constants in physics.

$$F_g = mg = \frac{Gm_1 m_2}{r^2} \rightarrow g = \frac{Gm_{Earth}}{r^2}$$

$$g = \frac{(6.67 \times 10^{-11} \, ^{N \cdot m^2}/_{kg^2})(5.98 \times 10^{24} \, kg)}{(6.37 \times 10^6 \, m + 380 \times 10^3 \, m)^2}$$

$$g = 8.75 \, ^{N}/_{kg} = 8.75 \, ^{m}/_{s^2}$$

This means that the acceleration due to gravity at the altitude the astronauts are orbiting the earth is only 11% less than on the surface of the Earth! In actuality, the Space Shuttle is falling, but it's moving so fast horizontally that by the time it falls, the Earth has curved away underneath it so that the shuttle remains at the same distance from the center of the Earth. It is in orbit! Of course, this takes tremendous speeds. To maintain an orbit of 380 km, the space shuttle travels approximately 7680 m/s, more than 23 times the speed of sound at sea level!

Geosynchronous orbit occurs when a satellite maintains an orbit above the same point on the Earth (the satellite's rotational velocity matches the Earth's rotational velocity). To "speed up" or "slow down" an orbiting satellite relative to the surface of the Earth, you just change the satellite's altitude.

5.30 Q: Calculate the magnitude of the centripetal force acting on Earth as it orbits the Sun, assuming a circular orbit and an orbital speed of 3.00 × 10⁴ meters per second.

5.30 A: Look up the mass of the Earth and the radius of the Earth's orbit.

$$F_c = ma_c = m \frac{v^2}{r}$$

$$F_c = (5.98 \times 10^{24} \, kg) \frac{(3 \times 10^4 \, ^{m}/_s)^2}{(1.5 \times 10^{11} \, m)} = 3.59 \times 10^{22} \, N$$

5.31 Q: The diagram below represents two satellites of equal mass, A and B, in circular orbits around a planet.

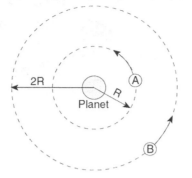

Compared to the magnitude of the gravitational force of attraction between satellite A and the planet, the magnitude of the gravitational force of attraction between satellite B and the planet is

(1) half as great

(2) twice as great

(3) one-fourth as great

(4) four times as great

5.31 A: (3) one-fourth as great due to the inverse square law relationship.

Kepler's Laws of Planetary Motion

In the early 1600s, most of the scientific world believed that the planets should have circular orbits, and many believed that the Earth was the center of the solar system. Using data collected by Tycho Brahe, German astronomer Johannes Kepler developed three laws governing the motion of planetary bodies, which described their orbits as ellipses with the sun at one of the focal points (even though the orbits of many planets are nearly circular). These laws are known as Kepler's Laws of Planetary Motion.

Kepler's First Law of Planetary Motion states that the orbits of planetary bodies are ellipses with the sun at one of the two foci of the ellipse.

Kepler's Second Law of Planetary Motion states that if you were to draw a line from the sun to the orbiting body, the body would sweep out equal areas along the ellipse in equal amounts of time. This is easier to observe graphically. In the diagram, if the orbiting body moves from point 1 to point 2 in the same amount of time as it moves from point 3 to point 4, then areas A1 and A2 must also be equal.

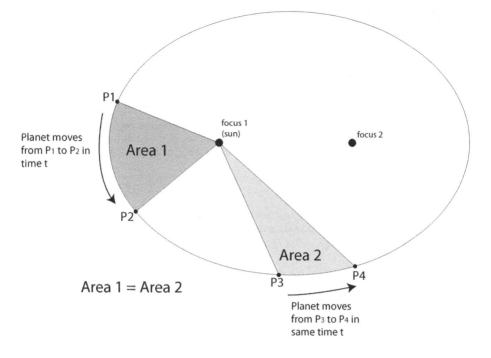

Kepler's 3rd Law of Planetary Motion, described several years after the first two laws were published, states that the ratio of the squares of the periods of two planets is equal to the ratio of the cubes of their semi-major axes. If this sounds confusing, don't worry, once again it's not as bad as it looks. The semi-major axis is the distance of the planet from the sun. What this is really saying, then, is that planets that are closer to the sun (with a smaller semi-major axis) have much shorter periods than planets that are farther from the sun. For example, the planet Mercury, closest to the sun, has an orbital period of 88 days. Neptune, which is 30 times farther from the sun than Earth, has an orbital period of 165 Earth years.

5.32 Q: Given the elliptical planetary orbit shown above, identify the interval during which the planet travels with the highest speed.

(1) Interval P1 to P2

(2) Interval P3 to P4

(3) They are the same.

5.32 A: (1) Interval P1 to P2. Because Area 1 is equal to Area 2, we know that the time interval from P1 to P2 must be equal to the time interval from P3 to P4 by Kepler's 2nd Law of Planetary Motion. Since the planet travels a greater distance from P1 to P2, it must have the higher speed during this portion of its journey.

5.33 Q: The shape of Mars' orbit around the sun is most accurately described as a:

(1) circle

(2) ellipse

(3) parabola

(4) hyperbola

5.33 A: (2) ellipse. The orbits of planets are ellipses with the sun as one of the foci of the ellipse. Note that even though the orbits are best described as ellipses, many of the planetary orbits are "nearly circular."

5.34 Q: Which planet takes the longest amount of time to make one complete revolution around the sun?

(1) Venus

(2) Earth

(3) Jupiter

(4) Uranus

5.34 A: (4) Uranus must have the longest orbital period since it is farthest from the sun according to Kepler's 3rd Law of Planetary Motion.

Chapter 6: Rotational Motion

"I shall now recall to mind that the motion of the heavenly bodies is circular, since the motion appropriate to a sphere is rotation in a circle."

— Nicolaus Copernicus

Objectives

1. Differentiate between translational and rotational motion of an object.
2. Describe the rotational motion of an object in terms of rotational position, velocity, and acceleration.
3. Use rotational kinematic equations to solve problems for objects rotating at constant acceleration.
4. Utilize the definitions of torque and Newton's 2nd Law for Rotational Motion to solve static equilibrium problems.
5. Explain what is meant by conservation of angular momentum.
6. Calculate the rotational kinetic energy and total kinetic energy of a rotating object moving through space.

The motion of objects cannot always be described completely using the laws of physics you've looked at so far. Besides motion, which changes an object's overall position (translational motion, or translational displacement), many objects also rotate around an axis, known as rotational, or angular, motion. The motion of some objects involves both translational and rotational motion.

An arrow speeding to its target, a hovercraft maneuvering through a swamp, and a hot air balloon floating through the atmosphere all experience only translational motion. A Ferris Wheel at an amusement park, a top spinning on a table, and a carousel at the beach experience only rotational motion. The Earth rotating around its axis (rotational motion) and moving through space (translational motion), and a frisbee spinning around its center while also flying through the air, both demonstrate simultaneous translational and rotational motion.

Rotational Kinematics

Typically people discuss rotational motion in terms of degrees, where one entire rotation around a circle is equal to 360°. When dealing with rotational motion from a physics perspective, measuring rotational motion in units known as radians (rads) is much more efficient. A radian measures a distance around an arc equal to the length of the arc's radius.

Up to this point, you've described distances and displacements in terms of Δx and Δy. In discussing angular displacements, you must transition to describing the translational displacement around an arc in terms of the variable s, while continuing to use the symbol θ (theta) to represent angles and angular displacement.

The distance completely around a circular path (360°), known as the circumference, C, can be found using $\Delta s = C = 2\pi r = 2\pi$ radians. Therefore, you can use this as a conversion factor to move back and forth between degrees and radians.

6.1 Q: Convert 90° to radians.

6.1 A: $90° \times \dfrac{2\pi \text{ rads}}{360°} = \dfrac{\pi}{2} \text{ rads} = 1.57 \text{ rads}$

6.2 Q: Convert 6 radians to degrees.

6.2 A: $6 \text{ rads} \times \dfrac{360°}{2\pi \text{ rads}} = 344°$

Angles are also measured in terms of revolutions (complete trips around a circle). A complete single rotation is equal to 360°, therefore you can write the conversion factors for rotational distances and displacements as $360° = 2\pi \text{ radians} = 1 \text{ revolution}$.

6.3 Q: Convert 1.5 revolutions to both radians and degrees.

6.3 A: $1.5 \text{ revs} \times \dfrac{2\pi \text{ rad}}{1 \text{ rev}} = 3\pi \text{ rads}$

$1.5 \text{ revs} \times \dfrac{360°}{1 \text{ rev}} = 540°$

Rotational kinematics is extremely similar to translational kinematics, all you have to do is learn the rotational versions of the kinematic variables and equations. When you learned translation kinematics, displacement was discussed in terms of Δx. With rotational kinematics, you'll use the angular coordinate θ instead. When average velocity was introduced in the translational world, you used the formula:

$$\bar{v} = \frac{x - x_0}{t} = \frac{\Delta x}{t}$$

When exploring rotational motion, you'll talk about the angular velocity ω (omega), given in units of radians per second (rad/s). Because angular (rotational) velocity is a vector, define the positive direction of rotation as counter-clockwise around the circular path, and the negative direction as clockwise around the path.

$$\bar{\omega} = \frac{\theta - \theta_0}{t} = \frac{\Delta \theta}{t}$$

NOTE: Formally, the direction of angular vectors is determined by the right-hand rule. Wrap the fingers of your right hand in the direction of the rotational displacement, velocity, or acceleration, and your thumb will point in the vector's direction.

6.4 Q: A record spins on a phonograph at 33 rpms (revolutions per minute) clockwise. Find the angular velocity of the record.

6.4 A:

$$\bar{\omega} = \frac{\Delta\theta}{t} = \frac{-33 \text{ revs}}{1\min}$$

$$\frac{-33 \text{ revs}}{1\min} \times \frac{1\min}{60s} \times \frac{2\pi \text{ rad}}{1 \text{ rev}} = -3.46\,^{rad}\!/_{s}$$

Note that the angular velocity vector is negative because the record is rotating in a clockwise direction.

6.5 Q: Find the magnitude of Earth's angular velocity in radians per second.

6.5 A: Realizing that the Earth makes one complete revolution every 24 hours, we can estimate the magnitude of the Earth's angular velocity as:

$$\bar{\omega} = \frac{\Delta\theta}{t} = \frac{2\pi \text{ rads}}{24 \text{ hr}} \times \frac{1 \text{ hr}}{3600s} = 7.27 \times 10^{-5}\,^{rads}\!/_{s}$$

In similar fashion, when you learned about translational acceleration, you found acceleration as the rate of change of an object's translational velocity:

$$a = \frac{\Delta v}{t} = \frac{v - v_0}{t}$$

Chapter 6: Rotational Motion

Angular acceleration α (alpha), given in units of radians per second^2, is the rate of change of an object's angular velocity. Since angular acceleration is a vector as well, you can define its direction as positive for increasing angular velocities in the counter-clockwise direction, and negative for increasing angular velocities in the clockwise direction.

$$\alpha = \frac{\Delta \omega}{t}$$

6.6 Q: A frog rides a unicycle. If the unicycle wheel begins at rest, and accelerates uniformly in a clockwise direction to an angular velocity of 15 rpms in a time of 6 seconds, find the angular acceleration of the unicycle wheel.

6.6 A: First, convert 15 rpms to rads/s.

$$\frac{15 \text{ revs}}{\min} \times \frac{1 \min}{60s} \times \frac{2\pi \text{ rad}}{1 \text{ rev}} = 1.57 \, ^{rad}\!/_{s}$$

Next, use the definition of angular acceleration.

$$\alpha = \frac{\Delta \omega}{t} = \frac{\omega - \omega_0}{t} = \frac{(1.57 \, ^{rad}\!/_{s})\text{-}0}{6s} = 0.26 \, ^{rad}\!/_{s^2}$$

Again, note the positive angular acceleration, as the bicycle wheel is accelerating in the counter-clockwise direction.

Putting these definitions together, you observe a very strong parallel between translational kinematic quantities and rotational kinematic quantities.

Variable	Translational	Angular
Displacement	Δs	$\Delta \theta$
Velocity	v	ω
Acceleration	a	α
Time	t	t

It's quite straightforward to translate between translational and angular variables as well when you know the radius (r) of the point of interest on a rotating object.

Variable	Translational	Angular
Displacement	$s = r\theta$	$\theta = \dfrac{s}{r}$
Velocity	$v = r\omega$	$\omega = \dfrac{v}{r}$
Acceleration	$a = r\alpha$	$\alpha = \dfrac{a}{r}$
Time	t	t

6.7 Q: A knight swings a mace of radius 1m in two complete revolutions. What is the translational displacement of the mace?

(1) 3.1 m
(2) 6.3 m
(3) 12.6 m
(4) 720 m

6.7 A: (3) $s = r\theta = (1m)(4\pi \text{ rads}) = 12.6m$

6.8 Q: A compact disc player is designed to vary the disc's rotational velocity so that the point being read by the laser moves at a linear velocity of 1.25 m/s. What is the CD's rotational velocity in revs/s when the laser is reading information on an inner portion of the disc at a radius of 0.03m?

6.8 A: $\omega = \dfrac{v}{r} = \dfrac{1.25\,^m\!/_s}{0.03m} = 41.7\,^{rad}\!/_s$

$\dfrac{41.7 \text{ rad}}{s} \times \dfrac{1 \text{ rev}}{2\pi \text{ rad}} = 6.63\,^{rev}\!/_s$

6.9 Q: What is the rotational velocity of the compact disc in the previous problem when the laser is reading the outermost portion of the disc (radius=0.06m)?

6.9 A: $\omega = \dfrac{v}{r} = \dfrac{1.25\,^m/_s}{0.06m} = 20.8\,^{rad}/_s$

$\dfrac{20.8\ \text{rad}}{s} \times \dfrac{1\ \text{rev}}{2\pi\ \text{rad}} = 3.32\,^{rev}/_s$

6.10 Q: A carousel accelerates from rest to an angular velocity of 0.3 rad/s in a time of 10 seconds. What is its angular acceleration? What is the linear acceleration for a point at the outer edge of the carousel, at a radius of 2.5 meters from the axis of rotation?

6.10 A: $\alpha = \dfrac{\omega - \omega_0}{t} = \dfrac{0.3\,^{rad}/_s}{10s} = 0.03\,^{rad}/_{s^2}$

$a = r\alpha = (2.5m)(0.03\,^{rad}/_{s^2}) = 0.075\,^m/_{s^2}$

The parallels between translational and rotational motion go even further. You developed a set of kinematic equations for translational motion that allowed you to explore the relationship between displacement, velocity, and acceleration. You can develop a corresponding set of relationships for angular displacement, angular velocity, and angular acceleration. The equations follow the same form as the translational equations, all you have to do is replace the translational variables with rotational variables, as shown in the following table.

Translational	Rotational
$v = v_0 + at$	$\omega = \omega_0 + \alpha t$
$\Delta x = v_0 t + \frac{1}{2}at^2$	$\Delta\theta = \omega_0 t + \frac{1}{2}\alpha t^2$
$v^2 = v_0^2 + 2a\Delta x$	$\omega^2 = \omega_0^2 + 2\alpha\Delta\theta$

The rotational kinematic equations can be used the same way you used the translational kinematic equations to solve problems. Once you know three of the kinematic variables, you can always use the equations to solve for the other two.

6.11 Q: A carpenter cuts a piece of wood with a high powered circular saw. The saw blade accelerates from rest with an angular acceleration of 14 rad/s² to a maximum speed of 15,000 rpms. What is the maximum speed of the saw in radians per second?

6.11 A:

$$\frac{15,000 \text{ revs}}{\text{min}} \times \frac{1 \text{ min}}{60s} \times \frac{2\pi \text{ rad}}{1 \text{ rev}} = 1570 \,^{rad}/_{s}$$

6.12 Q: How long does it take the saw to reach its maximum speed?

6.12 A: You can use our rotational kinematic equations to solve this problem:

Variable	Value
ω_0	0 rad/s
ω	1570 rad/s
$\Delta\theta$	?
α	14 rad/s²
t	FIND

$$\omega = \omega_0 + \alpha t$$

$$t = \frac{\omega - \omega_0}{\alpha} = \frac{1570 \,^{rad}/_{s} - 0}{14 \,^{rad}/_{s^2}} = 112s$$

6.13 Q: How many complete rotations does the saw make while accelerating to its maximum speed?

6.13 A: $\Delta\theta = \omega_0 t + \frac{1}{2}\alpha t^2$

$\Delta\theta = \frac{1}{2}(14\,{}^{rad}\!/_{s^2})(112s)^2 = 87{,}800$ rads

$87{,}800 \text{ rads} \times \dfrac{1 \text{ rev}}{2\pi \text{ rads}} = 14{,}000 \text{ revolutions}$

6.14 Q: A safety mechanism will bring the saw blade to rest in 0.3 seconds should the carpenter's hand come off the saw controls. What angular acceleration does this require? How many complete revolutions will the saw blade make in this time?

6.14 A: Begin by re-creating the rotational kinematics table.

Variable	Value
ω_0	1570 rad/s
ω	0 rad/s
$\Delta\theta$	FIND
α	FIND
t	0.3s

First, find the angular acceleration.

$\omega = \omega_0 + \alpha t$

$\alpha = \dfrac{\omega - \omega_0}{t} = \dfrac{0 - 1570\,{}^{rad}\!/_{s}}{0.3s} = -5230\,{}^{rad}\!/_{s^2}$

Next, find the angular displacement.

$\Delta\theta = \omega_0 t + \frac{1}{2}\alpha t^2$

$\Delta\theta = (1570\,{}^{rad}\!/_{s})(0.3s) + \frac{1}{2}(-5230\,{}^{rad}\!/_{s^2})(0.3s)^2$

$\Delta\theta = 236 \text{ rads}$

Finally, convert the angular displacement into revolutions

$\Delta\theta = 236 \text{ rads} \times \dfrac{1 \text{ rev}}{2\pi \text{ rads}} = 37.5 \text{ revolutions}$

Torque

Torque (τ) is a force that causes an object to turn. If you think about using a wrench to tighten a bolt, the closer to the bolt you apply the force, the harder it is to turn the wrench, while the farther from the bolt you apply the force, the easier it is to turn the wrench. This is because you generate a larger torque when you apply a force at a greater distance from the axis of rotation.

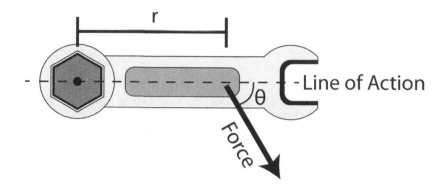

Let's take a look at the example of a wrench turning a bolt. A force is applied at a distance from the axis of rotation. Call this distance r. When you apply forces at 90 degrees to the imaginary line leading from the axis of rotation to the point where the force is applied (known as the line of action), you obtain maximum torque. As the angle at which the force applied decreases (θ), so does the torque causing the bolt to turn. Therefore, you can calculate the torque applied as:

$$\tau = Fr\sin\theta$$

In some cases, physicists will refer to the rsinθ as the lever arm, or moment arm, of the system. The lever arm is the perpendicular distance from the axis of rotation to the point where the force is applied. Alternately, you could think of torque as the component of the force perpendicular to the lever multiplied by the distance r. Units of torque are the units of force × distance, or Newton-meters (N·m).

6.15 Q: A pirate captain takes the helm and turns the wheel of his ship by applying a force of 20 Newtons to a wheel spoke. If he applies the force at a radius of 0.2 meters from the axis of rotation, at an angle of 80° to the line of action, what torque does he apply to the wheel?

6.15 A: $\tau = Fr\sin\theta$

$\tau = (20\,N)(0.2\,m)\sin(80°) = 3.94\,N \bullet m$

6.16 Q: A mechanic tightens the lugs on a tire by applying a torque of 110 N·m at an angle of 90° to the line of action. What force is applied if the wrench is 0.4 meters long?

6.16 A: $\tau = Fr\sin\theta$

$$F = \frac{\tau}{r\sin\theta} = \frac{110N \bullet m}{(0.4m)\sin 90°} = 275N$$

6.17 Q: How long must the wrench be if the mechanic is only capable of applying a force of 200N?

6.17 A: $\tau = Fr\sin\theta$

$$r = \frac{\tau}{F\sin\theta} = \frac{110N \bullet m}{(200N)\sin 90°} = .55m$$

Objects which have no rotational acceleration, or a net torque of zero, are said be in rotational equilibrium. This implies that any net positive (counter-clockwise) torque is balanced by an equal net negative (clockwise) torque.

Moment of Inertia

Previously, the inertial mass of an object (its translational inertia) was defined as that object's ability to resist a linear acceleration. Similarly, an object's rotational inertia, or **moment of inertia**, describes an object's resistance to a rotational acceleration. The symbol for an object's moment of inertia is I.

Objects that have most of their mass near their axis of rotation have a small rotational inertia, while objects that have more mass farther from the axis of rotation have larger rotational inertias.

For common objects, you can look up the formula for their moment of inertia. For more complex objects, the moment of inertia can be calculated by taking the sum of all the individual particles of mass making up the object multiplied by the square of their radius from the axis of rotation. This can be quite cumbersome using algebra, and is therefore typically left to calculus-based courses or numerical approximation using computing systems.

Commonly Used Moments of Inertia

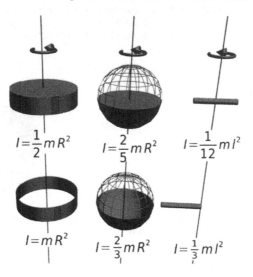

$$I = \frac{1}{2}mR^2 \qquad I = \frac{2}{5}mR^2 \qquad I = \frac{1}{12}ml^2$$

$$I = mR^2 \qquad I = \frac{2}{3}mR^2 \qquad I = \frac{1}{3}ml^2$$

6.18 Q: Calculate the moment of inertia for a solid sphere with a mass of 10 kg and a radius of 0.2m.

6.18 A: $I = \frac{2}{5}mR^2$

$I = \frac{2}{5}(10kg)(0.2m)^2$

$I = 0.16kg \bullet m^2$

6.19 Q: Calculate the moment of inertia for a hollow sphere with a mass of 10 kg and a radius of 0.2 m.

6.19 A: $I = \frac{2}{3}mR^2$

$I = \frac{2}{3}(10kg)(0.2m)^2$

$I = 0.27kg \bullet m^2$

6.20 Q: Calculate the moment of inertia for a long thin rod with a mass of 2 kg and a length of 1m rotating around the center of its length.

6.20 A: $I = \frac{1}{12}ml^2$

$I = \frac{1}{12}(2kg)(1m)^2$

$I = 0.17kg \bullet m^2$

6.21 Q: Calculate the moment of inertia for a long thin road with a mass of 2kg and a length of 1m rotating about its end.

6.21 A: $I = \frac{1}{3}ml^2$

$I = \frac{1}{3}(2kg)(1m)^2$

$I = 0.67kg \bullet m^2$

Newton's 2nd Law for Rotation

In the chapter on dynamics, you learned about forces causing objects to accelerate. The larger the net force, the greater the linear (or translational) acceleration, and the larger the mass of the object, the smaller the translational acceleration.

$$F_{net} = ma$$

The rotational equivalent of this law, Newton's 2nd Law for Rotation, relates the torque on an object to its resulting angular acceleration. The larger the net torque, the greater the rotational acceleration, and the larger the rotational inertia, the smaller the rotational acceleration:

$$\tau_{net} = I\alpha$$

6.22 Q: What is the angular acceleration experienced by a uniform solid disc of mass 2 kg and radius 0.1 m when a net torque of 10 N·m is applied? Assume the disc spins about its center.

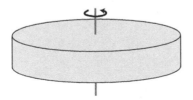

6.22 A: $\tau_{net} = I\alpha = \frac{1}{2}mR^2\alpha$

$\alpha = \frac{2\tau_{net}}{mR^2} = \frac{2\times10N \bullet m}{(2kg)(0.1m)^2} = 1000\,{}^{rad}\!/_{s^2}$

6.23 Q: A Round-A-Bout on a playground with a moment of inertia of 100 kg·m² starts at rest and is accelerated by a force of 150N at a radius of 1m from its center.

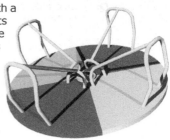

If this force is applied at an angle of 90° from the line of action for a time of 0.5 seconds, what is the final rotational velocity of the Round-A-Bout?

6.23 A: Start by making our rotational kinematics table:

Variable	Value
ω_0	0 rad/s
ω	FIND
$\Delta\theta$	?
α	?
t	0.5s

Since you only know two items on the table, you must find a third before you solve this with the rotational kinematic equations. Since you are given the moment of inertia of the Round-A-Bout as well as the applied force, you can solve for the angular acceleration using Newton's 2nd Law for Rotational Motion.

$$\tau_{net} = I\alpha$$

$$\alpha = \frac{\tau_{net}}{I} = \frac{Fr\sin\theta}{I} = \frac{(150N)(1m)\sin 90°}{100kg \bullet m^2} = 1.5\,{}^{rad}\!/\!_{s^2}$$

Now, use your rotational kinematics to solve for the final angular velocity of the Round-A-Bout.

$$\omega = \omega_0 + \alpha t$$

$$\omega = 0 + (1.5\,{}^{rad}\!/\!_{s^2})(0.5s) = 0.75\,{}^{rad}\!/\!_{s}$$

Angular Momentum

Previously, you learned that linear momentum, the product of an object's mass and its velocity, is conserved in a closed system. In similar fashion, spin angular momentum L, the product of an object's moment of inertia and its angular velocity about the center of mass, is also conserved in a closed system with no external net torques applied.

$$L = I\omega$$

This can be observed by watching a spinning ice skater. As an ice skater launches into a spin, she generates rotational velocity by applying a torque to her body. The skater now has an angular momentum as she spins around an axis which is equal to the product of her moment of inertia (rotational inertia) and her rotational velocity.

To increase the rotational velocity of her spin, she pulls her arms in close to her body, reducing her moment of inertia. Angular momentum is conserved, therefore rotational velocity must increase. Then, before coming out of the spin, the skater reduces her rotational velocity by move her arms away from her body, increasing her moment of inertia.

6.24 Q: Angelina spins on a rotating pedestal with an angular velocity of 8 radians per second. Bob throws her an exercise ball, which increases her moment of inertia from 2 kg·m² to 2.5 kg·m². What is Angelina's angular velocity after catching the exercise ball? (Neglect any external torque from the ball.)

6.24 A: Since there are no external torques, you know that the initial spin angular momentum must equal the final spin angular momentum, and can therefore solve for Angelina's final angular velocity:

$$L_0 = L \rightarrow I_0 \omega_0 = I\omega$$

$$\omega = \frac{I_0 \omega_0}{I} = \frac{(2kg \bullet m^2)(8\,{}^{rad}\!/_s)}{2.5kg \bullet m^2} = 6.4\,{}^{rad}\!/_s$$

6.25 Q: A disc with moment of inertia 1 kg·m² spins about an axle through its center of mass with angular velocity 10 rad/s. An identical disc which is not rotating is slid along the axle until it makes contact with the first disc. If the two discs stick together, what is their combined angular velocity?

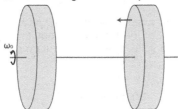

6.25 A: Once again, since there are no external torques, you know that spin angular momentum must be conserved. When the two discs stick together, their new combined moment of inertia must be the sum of their individual moments of inertia, for a total moment of inertia of 2 kg·m².

$$L_0 = L \rightarrow I_0 \omega_0 = I\omega$$

$$\omega = \frac{I_0 \omega_0}{I} = \frac{(1kg \bullet m^2)(10\,{}^{rad}\!/_s)}{2kg \bullet m^2} = 5\,{}^{rad}\!/_s$$

Rotational Kinetic Energy

When kinematics was first introduced, kinetic energy was defined as the ability of a moving object to move another object. Then, translational kinetic energy for a moving object was calculated using the formula:

$$KE = \tfrac{1}{2}mv^2$$

Since an object which is rotating also has the ability to move another object, it, too, must have kinetic energy. Rotational kinetic energy can be calculated using the analog to the translational kinetic energy formula -- all you have to do is replace inertial mass with moment of inertia, and translational velocity with angular velocity!

$$KE_{rot} = \tfrac{1}{2}I\omega^2$$

If an object exhibits both translational motion and rotational motion, the total kinetic energy of the object can be found by adding the translational kinetic energy and the rotational kinetic energy:

$$KE = \tfrac{1}{2}mv^2 + \tfrac{1}{2}I\omega^2$$

Because you're solving for energy, of course the answers will have units of Joules.

6.26 Q: Gina rolls a bowling ball of mass 7 kg and radius 10.9 cm down a lane with a velocity of 6 m/s. Find the rotational kinetic energy of the bowling ball, assuming it does not slip.

6.26 A: To find the rotational kinetic energy of the bowling ball, you need to know its moment of inertia and its angular velocity. Assume the bowling ball is a solid sphere to find its moment of inertia.

$$I = \tfrac{2}{5}mR^2$$

$$I = \tfrac{2}{5}(7kg)(.109m)^2 = 0.0333kg \bullet m^2$$

Next, find the ball's angular velocity.

$$\omega = \frac{v}{r} = \frac{6\,^m/_s}{.109m} = 55\,^{rad}/_s$$

Finally, solve for the rotational kinetic energy of the bowling ball.

$$KE_{rot} = \tfrac{1}{2}I\omega^2 = \tfrac{1}{2}(0.0333kg \bullet m^2)(55\,^{rad}/_s)^2 = 50.4J$$

6.27 Q: Find the total kinetic energy of the bowling ball from the previous problem.

6.27 A: The total kinetic energy is the sum of the translational kinetic energy and the rotational kinetic energy of the bowling ball.

$$KE = \tfrac{1}{2}mv^2 + \tfrac{1}{2}I\omega^2$$

$$KE = \tfrac{1}{2}(7kg)(6\,{}^m\!/_s)^2 + 50.4J$$

$$KE = 176J$$

6.28 Q: Harrison kicks a soccer ball which rolls across a field with a velocity of 5 m/s. What is the ball's total kinetic energy? You may assume the ball doesn't slip, and treat it as a hollow sphere of mass 0.43 kg and radius 0.11 meter.

6.28 A: Immediately note that the ball will have both translational and rotational kinetic energy. Therefore, you'll need to know the ball's mass (given), translational velocity (given), moment of inertia (unknown), and rotational velocity (unknown).

Start by finding the moment of inertia of the ball, modeled as a hollow sphere.

$$I = \tfrac{2}{3}mR^2$$

$$I = \tfrac{2}{3}(0.43kg)(0.11m)^2 = 0.00347kg \bullet m^2$$

Next, find the rotational velocity of the soccer ball.

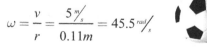

$$\omega = \frac{v}{r} = \frac{5\,{}^m\!/_s}{0.11m} = 45.5\,{}^{rad}\!/_s$$

Now you have enough information to calculate the total kinetic energy of the soccer ball.

$$KE = \tfrac{1}{2}mv^2 + \tfrac{1}{2}I\omega^2$$

$$KE = \tfrac{1}{2}(0.43kg)(5\,{}^m\!/_s)^2 + \tfrac{1}{2}(0.00347kg \bullet m^2)(45.5\,{}^{rad}\!/_s)^2$$

$$KE = 8.96J$$

Putting all this information together, rotational physics mirrors translational physics in terms of both variables and formulas. These equivalencies and relationships are summarized below.

Variable	Translational	Angular
Displacement	Δs	Δθ
Velocity	v	ω
Acceleration	a	α
Time	t	t
Force/Torque	F	τ
Mass/Moment of Inertia	m	I

Variable	Translational	Angular
Displacement	$s = r\theta$	$\theta = \dfrac{s}{r}$
Velocity	$v = r\omega$	$\omega = \dfrac{v}{r}$
Acceleration	$a = r\alpha$	$\alpha = \dfrac{a}{r}$
Time	t	t
Force/Torque	$F_{net} = ma$	$\tau_{net} = I\alpha$
Momentum	$p = mv$	$L = I\omega$
Kinetic Energy	$KE = \tfrac{1}{2}mv^2$	$KE = \tfrac{1}{2}I\omega^2$

Chapter 7: Work, Energy & Power

"Ambition is like a vector; it needs magnitude and direction.
Otherwise, it's just energy."

— Grace Lindsay

Objectives

1. Define work and calculate the work done by a force.
2. Calculate the kinetic energy of a moving object.
3. Determine the gravitational potential energy of a system.
4. Calculate the power of a system.
5. Apply conservation of energy to analyze energy transitions and transformations in a system.
6. Analyze the relationship between work done on or by a system, and the energy gained or lost by that system.
7. Use Hooke's Law to determine the elastic force on an object.
8. Calculate a system's elastic potential energy.

Work, energy and power are highly inter-related concepts that come up regularly in everyday life. You do work on an object when you move it. The rate at which you do the work is your power output. When you do work on an object, you transfer energy from one object to another. In this chapter you'll explore how energy is transferred and transformed, how doing work on an object changes its energy, and how quickly work can be done.

Work

Sometimes you work hard. Sometimes you're a slacker. But, right now, are you doing work? And what is meant by the word "work?" In physics terms, **work** is the process of moving an object by applying a force.

I'm sure you can think up countless examples of work being done, but a few that spring to mind include pushing a snowblower to clear the driveway, pulling a sled up a hill with a rope, stacking boxes of books from the floor onto a shelf, and throwing a baseball from the pitcher's mound to home plate.

Let's take a look at a few scenarios and investigate what work is being done.

In the first scenario, a monkey in a jet pack blasts through the atmosphere, accelerating to higher and higher speeds. In this case, the jet pack is applying a force causing it to move. But what is doing the work? Hot expanding gases are pushed backward out of the jet pack. Using Newton's 3rd Law, you observe the reactionary force of the gas pushing the jet pack forward, causing a displacement. Therefore, the expanding exhaust gas is doing work on the jet pack.

In the second scenario, a girl struggles to push her stalled car, but can't make it move. Even though she's expending significant effort, no work is being done on the car because it isn't moving.

In the final scenario, a child in a ghost costume carries a bag of Halloween candy across the yard. In this situation, the child applies a force upward on the bag, but the bag moves horizontally. From this perspective, the forces of the child's arms on the bag don't cause the displacement, therefore no work is being done by the child.

Mathematically, work can be expressed by the following equation:

$$W = F\Delta r$$

W is the work done, **F** is the force applied in Newtons, and **Δr** is the object's displacement in meters.

The units of work can be found by performing unit analysis on the work formula. If work is force multiplied by distance, the units must be the units of force multiplied by the units of distance, or newtons multiplied by meters. A newton-meter is also known as a Joule (J).

It's important to note that when using this equation, only the force applied in the direction of the object's displacement counts! This means that if the force and displacement vectors aren't in exactly the same direction, you need to take the component of force in the direction of the object's displacement. To do this, line up the force and displacement vectors tail-to-tail and measure the angle between them. Since this component of force can be calculated by multiplying the force by the cosine of the angle between the force and displacement vectors, you can re-write the work equation as:

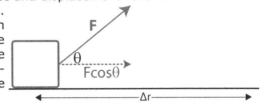

$$W = F\cos\theta \bullet \Delta r = Fr\cos\theta$$

7.01 Q: An appliance salesman pushes a refrigerator 2 meters across the floor by applying a force of 200N. Find the work done.

7.01 A: Since the force and displacement are in the same direction, the angle between them is 0.
$$W = F\Delta r\cos\theta = (200N)(2m)\cos0° = 400J$$

7.02 Q: A friend's car is stuck on the ice. You push down on the car to provide more friction for the tires (by way of increasing the normal force), allowing the car's tires to propel it forward 5m onto less slippery ground. How much work did you do?

7.02 A: You applied a downward force, yet the car's displacement was sideways. Therefore, the angle between the force and displacement vectors is 90°.
$$W = F\Delta r\cos\theta = F\Delta r\cos90° = 0$$

7.03 Q: You push a crate up a ramp with a force of 10N. Despite your pushing, however, the crate slides down the ramp a distance of 4m. How much work did you do?

7.03 A: Since the direction of the force you applied is opposite the direction of the crate's displacement, the angle between the two vectors is 180°.
$$W = F\Delta r\cos\theta = (10N)(4m)\cos180° = -40J$$

7.04 Q: How much work is done in lifting an 8-kg box from the floor to a height of 2m above the floor?

7.04 A: It's easy to see the displacement is 2m, and the force must be applied in the direction of the displacement, but what is the force? To lift the box you must match and overcome the force of gravity on the box. Therefore, the force applied is equal to the gravitational force, or weight, of the box, mg=(8kg)(9.81m/s²)=78.5N.

$$W = F\Delta r\cos\theta = (78.5N)(2m)\cos 0° = 157J$$

7.05 Q: Barry and Sidney pull a 30-kg wagon with a force of 500N a distance of 20m. The force acts at a 30° angle to the horizontal. Calculate the work done.

7.05 A:
$$W = F\Delta r\cos\theta = (500N)(20m)\cos 30° = 8660J$$

7.06 Q: The work done in lifting an apple one meter near Earth's surface is approximately

(1) 1 J

(2) 0.01 J

(3) 100 J

(4) 1000 J

7.06 A: (1) The trick in this problem is recalling the approximate weight of an apple. With an "order-of-magnitude" estimate, you can say an apple has a mass of 0.1 kg, or a weight of 1 N. Given this information, the work done is:

$$W = F\Delta r\cos\theta = (1N)(1m)\cos 0° = 1J$$

7.07 Q: As shown in the diagram, a child applies a constant 20-newton force along the handle of a wagon which makes a 25° angle with the horizontal.

How much work does the child do in moving the wagon a horizontal distance of 4.0 meters?

(1) 5.0 J

(2) 34 J

(3) 73 J

(4) 80. J

7.07 A: (4) $W = F\Delta r\cos\theta = (20N)(4m)\cos(25°) = 73J$

Force vs. Displacement Graphs

The area under a force vs. displacement graph is the work done by the force. Consider the situation of a block being pulled across a table with a constant force of 5 Newtons over a displacement of 5 meters, then the force gradually tapers off over the next 5 meters.

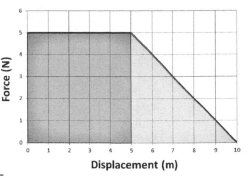

The work done by the force moving the block can be calculated by taking the area under the force vs. displacement graph (a combination of a rectangle and triangle) as follows:

$$Work = Area_{rectangle} + Area_{triangle}$$

$$Work = lw + \frac{1}{2}bh$$

$$Work = (5m)(5N) + \frac{1}{2}(5m)(5N)$$

$$Work = 37.5J$$

7.08 Q: A boy pushes his wagon at constant speed along a level sidewalk. The graph below represents the relationship between the horizontal force exerted by the boy and the distance the wagon moves.

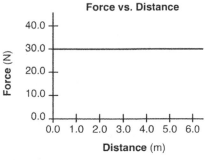

What is the total work done by the boy in pushing the wagon 4.0 meters?

(1) 5.0 J

(2) 7.5 J

(3) 120 J

(4) 180 J

7.08 A: (3) 120 J

$$Work = Area_{rectangle} = lw = (4m)(30N) = 120J$$

7.09 Q: A box is wheeled to the right with a varying horizontal force. The graph below represents the relationship between the applied force and the distance the box moves.

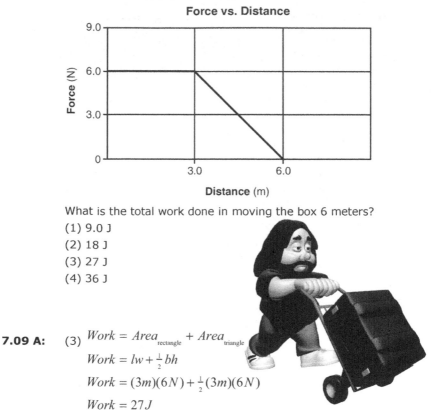

Force vs. Distance

What is the total work done in moving the box 6 meters?

(1) 9.0 J

(2) 18 J

(3) 27 J

(4) 36 J

7.09 A: (3) $Work = Area_{rectangle} + Area_{triangle}$

$Work = lw + \frac{1}{2}bh$

$Work = (3m)(6N) + \frac{1}{2}(3m)(6N)$

$Work = 27J$

Hooke's Law

An interesting application of work combined with the Force and Displacement graph is examining the force applied by a spring. The more you stretch a spring, the greater the force of the spring. Similarly, the more you compress a spring, the greater the force. This can be modeled as a linear relationship, where the force applied by the spring is equal to a constant multiplied by the displacement of the spring.

$$F_s = kx$$

F_s is the force of the spring in newtons, x is the displacement of the spring from its equilibrium (or rest) position, in meters, and k is the spring constant, which tells you how stiff or powerful a spring is, in newtons per meter. The larger the spring constant, k, the more force the spring applies per amount of displacement.

You can determine the spring constant of a spring by making a graph of the force from a spring on the y-axis, and placing the displacement of the spring from its equilibrium, or rest position, on the x-axis. The slope of the graph will give you the spring constant. For the case of the spring depicted in the graph at right, you can find the spring constant as follows:

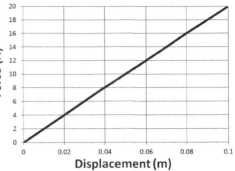

$$k = Slope = \frac{rise}{run} = \frac{\Delta F}{\Delta x} = \frac{20N - 0N}{0.1m - 0m} = 200\,{}^{N}\!/\!_{m}$$

You must have done work to compress or stretch the spring, since you applied a force and caused a displacement. You can find the work done in stretching or compressing a spring by taking the area under the graph. For the spring shown, to displace the spring 0.1m, you can find the work done as shown below:

$$Work = Area_{tri} = \tfrac{1}{2}bh = \tfrac{1}{2}(0.1m)(20N) = 1J$$

7.10 Q: In an experiment, a student applied various forces to a spring and measured the spring's corresponding elongation. The table below shows his data.

Force (newtons)	Elongation (meters)
0	0
1.0	0.30
3.0	0.67
4.0	1.00
5.0	1.30
6.0	1.50

Plot force versus elongation and draw the best-fit line. Then, using your graph, calculate the spring constant of the spring. [Show all work, including the equation and substitution with units.]

7.10 A:

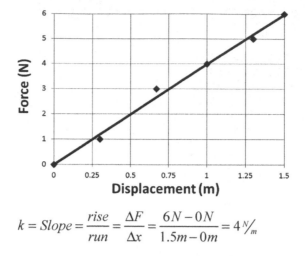

$$k = Slope = \frac{rise}{run} = \frac{\Delta F}{\Delta x} = \frac{6N - 0N}{1.5m - 0m} = 4\,^N\!/_m$$

7.11 Q: In a laboratory investigation, a student applied various downward forces to a vertical spring. The applied forces and the corresponding elongations of the spring from its equilibrium position are recorded in the data table.

Construct a graph, marking an appropriate scale on the axis labeled "Force (N)." Plot the data points for force versus elongation. Draw the best-fit line or curve. Then, using your graph, calculate the spring constant of this spring. [Show all work, including the equation and substitution with units.]

Force (newtons)	Elongation (meters)
0	0
0.5	0.010
1.0	0.018
1.5	0.027
2.0	0.035
2.5	0.046

7.11 A:

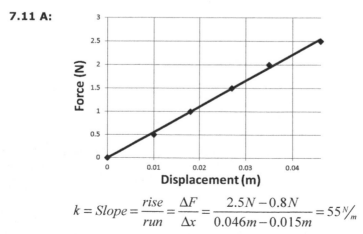

$$k = Slope = \frac{rise}{run} = \frac{\Delta F}{\Delta x} = \frac{2.5N - 0.8N}{0.046m - 0.015m} = 55\,^N\!/_m$$

7.12 Q: A 10-newton force compresses a spring 0.25 meter from its equilibrium position. Calculate the spring constant of this spring.

7.12 A: $F_s = kx$

$$k = \frac{F_s}{x} = \frac{10N}{0.25m} = 40 \, ^N\!/_m$$

Power

Power is a term used quite regularly in all aspects of life. People talk about how powerful the new boat motor is, the power of positive thinking, and even the power company's latest bill. All of these uses of the term power relate to how much work can be done in some amount of time.

In physics, work is the process of moving an object by applying a force. The rate at which the force does work is known as **power** (P). The units of power are the units of work divided by time, or Joules per second, known as a **Watt** (W).

$$P = \frac{W}{t}$$

Since power is the rate at which work is done, it is possible to have the same amount of work done but with a different supplied power, if the time is different.

7.13 Q: Rob and Peter move a sofa 3 meters across the floor by applying a combined force of 200N horizontally. If it takes them 6 seconds to move the sofa, what amount of power did they supply?

7.13 A: $P = \dfrac{W}{t} = \dfrac{F \Delta r \cos \theta}{t} = \dfrac{(200N)(3m)}{6s} = 100W$

7.14 Q: Kevin then pushes the same sofa 3 meters across the floor by applying a force of 200N. Kevin, however, takes 12 seconds to push the sofa. What amount of power did Kevin supply?

7.14 A: $P = \dfrac{W}{t} = \dfrac{F\Delta r \cos\theta}{t} = \dfrac{(200N)(3m)}{12s} = 50W$

As you can see, although Kevin did the same amount of work as Rob and Peter in pushing the sofa (600J), Rob and Peter supplied twice the power of Kevin because they did the same work in half the time!

There's more to the story, however. Since power is defined as work over time, and because work is equal to force (in the direction of the displacement) multiplied by displacement, you can replace work in the equation with F×r:

$$P = \dfrac{W}{t} = \dfrac{F\Delta r}{t}$$

Looking carefully at this equation, you can observe a displacement divided by time. Since displacement divided by time is the definition of average velocity, you can replace Δr/t with v in the equation to obtain:

$$P = \dfrac{W}{t} = \dfrac{F\Delta r}{t} = F\overline{v}$$

So, not only is power equal to work done divided by the time required, it's also equal to the force applied (in the direction of the displacement) multiplied by the average velocity of the object.

7.15 Q: Motor A lifts a 5000N steel crossbar upward at a constant 2 m/s. Motor B lifts a 4000N steel support upward at a constant 3 m/s. Which motor is supplying more power?

7.15 A: Motor B supplies more power than Motor A.

$P_{MotorA} = F\overline{v} = (5000N)(2\tfrac{m}{s}) = 10000W$

$P_{MotorB} = F\overline{v} = (4000N)(3\tfrac{m}{s}) = 12000W$

7.16 Q: A 70-kilogram cyclist develops 210 watts of power while pedaling at a constant velocity of 7 meters per second east. What average force is exerted eastward on the bicycle to maintain this constant speed?

(1) 490 N

(2) 30 N

(3) 3.0 N

(4) 0 N

7.16 A: (2) $P = F\overline{v}$

$$F = \frac{P}{\overline{v}} = \frac{210W}{7\,{}^m\!/_s} = 30N$$

7.17 Q: Alien A lifts a 500-newton child from the floor to a height of 0.40 meters in 2 seconds. Alien B lifts a 400-newton student from the floor to a height of 0.50 meters in 1 second. Compared to Alien A, Alien B does

(1) the same work but develops more power

(2) the same work but develops less power

(3) more work but develops less power

(4) less work but develops more power

7.17 A: (1) the same work but develops more power.

7.18 Q: A 110-kilogram bodybuilder and his 55-kilogram friend run up identical flights of stairs. The bodybuilder reaches the top in 4.0 seconds while his friend takes 2.0 seconds. Compared to the power developed by the bodybuilder while running up the stairs, the power developed by his friend is

(1) the same

(2) twice as much

(3) half as much

(4) four times as much

7.18 A: (1) the same.

Energy

We've all had days where we've had varying amounts of energy. You've gotten up in the morning, had to drag yourself out of bed, force yourself to get ready to school, and once you finally get to class, you don't have the energy to do much work. Other days, when you've had more energy, you may have woken up before the alarm clock, hustled to get ready for the day while a bunch of thoughts jump around in your head, and hurried on to begin your activities. Then, throughout the day, the more work you do, the more energy you lose... What's the difference in these days?

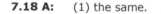

In physics, **energy** is the ability or capacity to do work. And as mentioned previously, work is the process of moving an object. So, if you combine the definitions, energy is the ability or capacity to move an object. So far you've examined kinetic energy, or energy of motion, and therefore kinetic energy must be the ability or capacity of a moving object to move another object! Mathematically, kinetic energy is calculated using the formula:

$$KE = \tfrac{1}{2}mv^2$$

Of course, there are more types of energy than just kinetic. Energy comes in many forms, which you can classify as kinetic (energy of motion) or potential (stored) to various degrees. This includes solar energy, thermal energy, gravitational potential energy, nuclear energy, chemical potential energy, sound energy, electrical energy, elastic potential energy, light energy, and so on. In all cases, energy can be transformed from one type to another and you can transfer energy from one object to another by doing work.

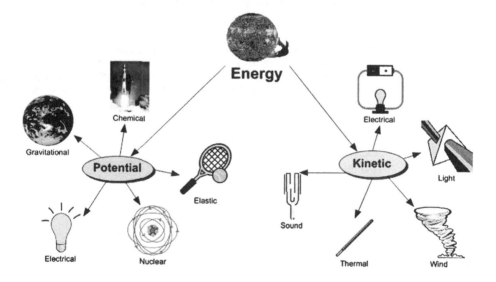

The units of energy are the same as the units of work, joules (J). Through dimensional analysis, observe that the units of KE (kg·m²/s²) must be equal to the units of work (N·m):

$$\frac{kg \bullet m^2}{s^2} = N \bullet m = J$$

7.19 Q: Which is an SI unit for work done on an object?

(1) $\dfrac{kg \bullet m^2}{s^2}$

(3) $\dfrac{kg \bullet m}{s}$

$$(2) \ \frac{kg \bullet m^2}{s} \qquad\qquad (4) \ \frac{kg \bullet m}{s^2}$$

7.19 A: (1) is equivalent to a newton-meter, also known as a Joule.

Gravitational Potential Energy

Potential energy is energy an object possesses due to its position or condition. **Gravitational potential energy** is the energy an object possesses because of its position in a gravitational field (height).

Assume a 10-kilogram box sits on the floor. You can arbitrarily call its current potential energy zero, just to give a reference point. If you do work to lift the box one meter off the floor, you need to overcome the force of gravity on the box (its weight) over a distance of one meter. Therefore, the work done on the box can be obtained from:

$$W = F\Delta r = (mg)h = (10kg)(9.8\,{}^m\!/_{s^2})(1m) = 98J$$

So, to raise the box to a height of 1m, you must do 98 Joules of work on the box. The work done in lifting the box is equal to the change in the potential energy of the box, so the box's gravitational potential energy must be 98 Joules.

When you performed work on the box, you transferred some of your stored energy to the box. Along the way, it just so happens that you derived the formula for the gravitational potential energy of an object. The change in the object's potential energy, ΔPE, is equal to the force of gravity on the box multiplied by its change in height, mgΔh.

$$\Delta PE = mg\Delta h$$

This formula can be used to solve a variety of problems involving the potential energy of an object.

7.20 Q: The diagram below represents a 155-newton box on a ramp. Applied force F causes the box to slide from point A to point B.

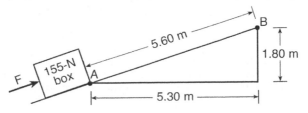

What is the total amount of gravitational potential energy gained by the box?

(1) 28.4 J

(2) 279 J

(3) 868 J

(4) 2740 J

7.20 A: (2) $\Delta PE = mg\Delta h = (155N)(1.8m) = 279J$

7.21 Q: Which situation describes a system with decreasing gravitational potential energy?

(1) a girl stretching a horizontal spring

(2) a bicyclist riding up a steep hill

(3) a rocket rising vertically from Earth

(4) a boy jumping down from a tree limb

7.21 A: (4) The boy's height above ground is decreasing, so his gravitational PE is decreasing.

7.22 Q: A car travels at constant speed v up a hill from point A to point B, as shown in the diagram below.

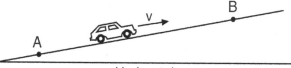

As the car travels from A to B, its gravitational potential energy

(1) increases and its kinetic energy decreases

(2) increases and its kinetic energy remains the same

(3) remains the same and its kinetic energy decreases

(4) remains the same and its kinetic energy remains the same

7.22 A: (2) The car's height above ground increases so gravitational potential energy increases, and velocity remains constant, so kinetic energy remains the same.

7.23 Q: An object is thrown vertically upward. Which pair of graphs best represents the object's kinetic energy and gravitational potential energy as functions of its displacement while it rises?

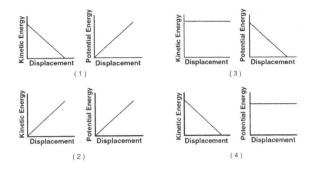

7.23 A: (1) shows the object's kinetic energy decreasing as it slows down on its way upward, while its potential energy increases as its height increases.

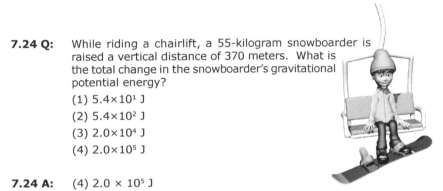

7.24 Q: While riding a chairlift, a 55-kilogram snowboarder is raised a vertical distance of 370 meters. What is the total change in the snowboarder's gravitational potential energy?

(1) 5.4×10^1 J

(2) 5.4×10^2 J

(3) 2.0×10^4 J

(4) 2.0×10^5 J

7.24 A: (4) 2.0×10^5 J

$$\Delta PE = mg\Delta h = (55kg)(9.8\,{}^m\!/_{s^2})(370m) = 2 \times 10^5 J$$

Elastic Potential Energy

Another form of potential energy involves the stored energy an object possesses due to its position in a stressed elastic system. An object at the end of a compressed spring, for example, has **elastic potential energy**. When the spring is released, the elastic potential energy of the spring will do work on the object, moving the object and transferring the energy of the spring into kinetic energy of the object. Other examples of elastic potential energy include tennis rackets, rubber bands, bows (as in bows and arrows), trampolines, bouncy balls, and even pole-vaulting poles.

The most common problems involving elastic potential energy in introductory physics involve the energy stored in a spring. As you learned in the previous topic on work, the force needed to compress or stretch a spring from its equilibrium position increases linearly. The more you stretch or compress the spring, the more force it applies trying to restore itself to its equilibrium position. This is called Hooke's Law:

$$F_s = kx$$

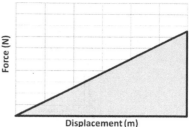

Force (N)

Displacement (m)

Further, you can find the work done in compressing or stretching the spring by taking the area under a force vs. displacement graph for the spring.

$$W = Fx = Area_{triangle} = \tfrac{1}{2}bh = \tfrac{1}{2}(x)(kx) = \tfrac{1}{2}kx^2$$

Since the work done in compressing or stretching the spring from its equilibrium position transfers energy to the spring, you can conclude that the potential energy stored in the spring must be equal to the work done to compress the spring. The potential energy of a spring is therefore given by:

$$PE_s = \tfrac{1}{2}kx^2$$

7.25 Q: A spring with a spring constant of 4.0 newtons per meter is compressed by a force of 1.2 newtons. What is the total elastic potential energy stored in this compressed spring?

(1) 0.18 J

(2) 0.36 J

(3) 0.60 J

(4) 4.8 J

7.25 A: PE_s can't be calculated directly since x isn't known, but x can be found from Hooke's Law:

$$F_s = kx$$

$$x = \frac{F_s}{k} = \frac{1.2N}{4\,\sfrac{N}{m}} = 0.3m$$

With x known, the potential energy equation for a spring can be utilized.

$$PE_s = \tfrac{1}{2}kx^2$$

$$PE_s = \tfrac{1}{2}(4\,\sfrac{N}{m})(0.3m)^2 = 0.18J$$

7.26 Q: An unstretched spring has a length of 10 centimeters. When the spring is stretched by a force of 16 newtons, its length is increased to 18 centimeters. What is the spring constant of this spring?

(1) 0.89 N/cm

(2) 2.0 N/cm

(3) 1.6 N/cm

(4) 1.8 N/cm

7.26 A: $F_s = kx$

$$k = \frac{F_s}{x} = \frac{16N}{8cm} = 2 \, ^N\!/_{cm}$$

7.27 Q: Which graph best represents the relationship between the elastic potential energy stored in a spring and its elongation from equilibrium?

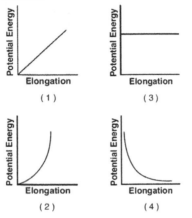

7.27 A: (2) due to the displacement^2 relationship.

7.28 Q: A pop-up toy has a mass of 0.020 kilogram and a spring constant of 150 newtons per meter. A force is applied to the toy to compress the spring 0.050 meter.

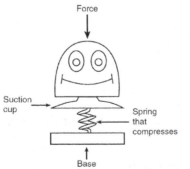

(A) Calculate the potential energy stored in the compressed spring.

(B) The toy is activated and all the compressed spring's potential energy is converted to gravitational potential energy. Calculate the maximum vertical height to which the toy is propelled.

7.28 A: (A) $PE_s = \frac{1}{2}kx^2$

$$PE_s = \frac{1}{2}(150\,{}^{N}\!/_{m})(0.05m)^2 = 0.1875J$$

(B) $PE_g = mgh$

$$h = \frac{PE_g}{mg} = \frac{0.1875J}{(0.02kg)(9.8\,{}^{m}\!/_{s^2})} = 0.96m$$

7.29 Q: A spring with a spring constant of 80 newtons per meter is displaced 0.30 meter from its equilibrium position. The potential energy stored in the spring is

(1) 3.6 J

(2) 7.2 J

(3) 12 J

(4) 24 J

7.29 A: (1) $PE_s = \frac{1}{2}kx^2 = \frac{1}{2}(80\,{}^{N}\!/_{m})(0.3m)^2 = 3.6J$

Work-Energy Theorem

Of course, there are many different kinds of energy which haven't been mentioned specifically. Energy can be converted among its many different forms, such as mechanical (which is kinetic, gravitational potential, and elastic potential), electromagnetic, nuclear, and thermal (or internal) energy.

When a force does work on a system, the work done changes the system's energy. If the work done increases motion, there is an increase in the system's kinetic energy. If the work done increases the object's height, there is an increase in the system's gravitational potential energy. If the work done compresses a spring, there is an increase in the system's elastic potential energy. If the work is done against friction, however, where does the energy go? In this case, the energy isn't lost, but instead increases the rate at which molecules in the object vibrate, increasing the object's temperature, or internal energy.

The understanding that the work done on a system by an external force changes the energy of the system is known as the Work-Energy Theorem. If an external force does positive work on the system, the system's total energy increases. If, instead, the system does work, the system's total energy decreases. Put another way, you add energy to a system by doing work on it and take energy from a system when the system does the work (much like you add value to your bank account by making a deposit and take value from your account by writing a check).

This relationship can be expressed by showing the formula for work as equal to the force times the displacement (F·Δr), as well as the change in total energy (ΔE):

$$W = F\Delta r = \Delta E_T$$

Sources of Energy on Earth

So where does all this energy initially come from? Here on Earth, the energy you deal with everyday ultimately comes from the conversion of mass into energy, the source of the sun's energy. The sun's radiation provides an energy source for life on earth, which over the millennia has become the source of fossil fuels. The sun's radiation also provides the thermal and light energy that heat the atmosphere and cause the winds to blow. The sun's energy evaporates water, which eventually recondenses as rain and snow, falling to the Earth's surface to create lakes and rivers, with gravitational potential energy, which is harnessed in hydroelectric power plants. Nuclear power also comes from the conversion of mass into energy. Just try to find an energy source on Earth that doesn't originate with the conversion of mass into energy!

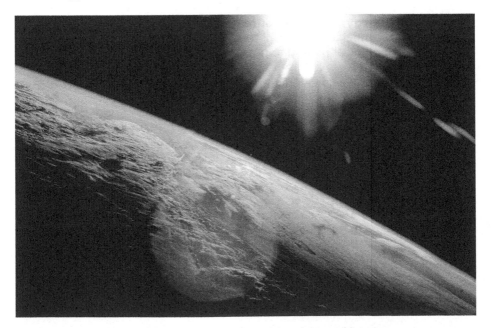

Conservation of Energy

"Energy cannot be created or destroyed... it can only be changed."

Chances are you've heard that phrase before. It's one of the most important concepts in all of physics. It doesn't mean that an object can't lose energy or gain energy. What it means is that energy can be changed into different forms, and transferred from system to system, but it never magically disappears or reappears.

Following up on the bank account analogy, if you have a certain amount of money in your bank account and then you spend some money, your bank account balance decreases. Your money wasn't lost, however, it was transferred to another system. It may even change forms... perhaps you purchased an item from another country. Your money is no longer in the form of dollars and cents, but is instead now part of another system in another currency.

There are some issues with the money analogy, however. If the total money supply in a country is low, a government can print more currency. In the world of physics, however, the total amount of energy throughout the universe is fixed. In other words, it cannot be replenished. Alternately, governments can collect and destroy currency -- in the world of physics, you can never truly destroy energy. The understanding that the total amount of energy in the universe remains fixed is known as the law of conservation of energy.

Mechanical energy is the sum of an object's kinetic energy as well as its gravitational potential and elastic potential energies. Non-mechanical energy forms include chemical potential, nuclear, and thermal.

Total energy is always conserved in any system, which is the law of conservation of energy. By confining the discussion to just the mechanical forms of energy, however, if you neglect the effects of friction you can also state that total mechanical energy is constant in any system.

Take the example of an F/A-18 Hornet jet fighter with a mass of 20,000 kilograms flying at an altitude of 10,000 meters above the surface of the earth with a velocity of 250 m/s. In this scenario, you can calculate the total mechanical energy of the jet fighter as follows:

$$E_T = PE_g + KE = mgh + \tfrac{1}{2}mv^2$$
$$E_T = (20000 kg)(9.8 \tfrac{m}{s^2})(10000 m) + \tfrac{1}{2}(20000 kg)(250 \tfrac{m}{s})^2$$
$$E_T = 2.59 \times 10^9 J$$

Now, assume the Hornet dives down to an altitude of 2,000 meters above the surface of the Earth. Total mechanical energy remains constant, and the gravitational potential energy of the fighter decreases, therefore the kinetic energy of the fighter must increase. The fighter's velocity goes up as a result of flying closer to the Earth! For this reason, a key concept in successful dogfighting taught to military pilots is that of energy conservation!

You can even calculate the new velocity of the fighter jet since you know its new height and its total mechanical energy must remain constant. Solving for velocity, you find that the Hornet has almost doubled its speed by "trading in" 8000 meters of altitude for velocity!

$$E_T = PE_g + KE = mgh + \frac{1}{2}mv^2$$

$$\frac{1}{2}mv^2 = E_T - mgh$$

$$v = \sqrt{\frac{2(E_T - mgh)}{m}}$$

$$v = \sqrt{\frac{2(2.59 \times 10^9 J - (20000kg)(9.8 \, ^m\!/_{s^2})(2000m))}{20000kg}}$$

$$v = 469 \, ^m\!/_s$$

If instead you had been told that some of the mechanical energy of the jet was lost to air resistance (friction), you could also account for that by stating that the total mechanical energy of the system is equal to the gravitational potential energy, the kinetic energy, and the change in internal energy of the system (Q). This leads to the conservation of mechanical energy formula:

$$E_T = PE + KE + Q$$

Let's take another look at free fall, only this time, you can analyze a falling object using the law of conservation of energy and compare it to the analysis using the kinematics equations studied previously.

The problem: An object falls from a height of 10m above the ground. Neglecting air resistance, find its velocity the moment before the object strikes the ground.

Conservation of Energy Approach: The energy of the object at its highest point must equal the energy of the object at its lowest point, therefore:

$$E_{top} = E_{bottom}$$

$$PE_{top} = KE_{bottom}$$

$$mgh_{top} = \frac{1}{2}mv^2_{bottom}$$

$$v_{bottom} = \sqrt{2gh} = \sqrt{2(9.8\,{}^m\!/\!_{s^2})(10m)} = 14\,{}^m\!/\!_s$$

Kinematics Approach: For an object in free fall, its initial velocity must be zero, its displacement is 10 meters, and the acceleration due to gravity on the surface of the Earth is 9.81 m/s². Choosing down as the positive direction:

Variable	Value
v_0	0 m/s
v	FIND
Δy	10 m
a	9.8 m/s²
t	?

$$v^2 = v_0^2 + 2a\Delta y$$

$$v = \sqrt{v_0^2 + 2a\Delta y} = \sqrt{2(9.8\,{}^m\!/\!_{s^2})(10m)} = 14\,{}^m\!/\!_s$$

As you can see, you reach the same conclusion regardless of approach!

7.30 Q: The diagram below shows a toy cart possessing 16 joules of kinetic energy traveling on a frictionless, horizontal surface toward a horizontal spring.

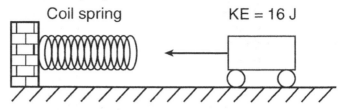

Coil spring   KE = 16 J

Frictionless, horizontal surface

If the cart comes to rest after compressing the spring a distance of 1.0 meter, what is the spring constant of the spring?

(1) 32 N/m

(2) 16 N/m

(3) 8.0 N/m

(4) 4.0 N/m

7.30 A: (1) $KE = PE_s = \frac{1}{2}kx^2$

$$k = \frac{2KE}{x^2} = \frac{2(16J)}{(1m)^2} = 32 \, \text{N/m}$$

7.31 Q: A child does 0.20 joules of work to compress the spring in a pop-up toy. If the mass of the toy is 0.010 kilograms, what is the maximum vertical height that the toy can reach after the spring is released?

(1) 20 m

(2) 2.0 m

(3) 0.20 m

(4) 0.020 m

7.31 A: (2) The potential energy in the compressed spring must be equal to the gravitational potential energy of the toy at its maximum vertical height.

$$PE_s = PE_g = mgh$$

$$h = \frac{PE_s}{mg} = \frac{0.2J}{(0.01kg)(9.8 \, \text{m/s}^2)} = 2m$$

7.32 Q: A lawyer knocks her folder of mass m off her desk of height h. What is the speed of the folder upon striking the floor?

(1) $\sqrt{(2gh)}$

(2) 2gh

(3) mgh

(4) mh

7.32 A: (1) The folder's initial gravitational energy becomes its kinetic energy right before striking the floor.

$$PE_{desk} = KE_{floor}$$

$$mgh = \tfrac{1}{2}mv^2$$

$$v = \sqrt{2gh}$$

7.33 Q: A 65-kilogram pole vaulter wishes to vault to a height of 5.5 meters.

(A) Calculate the minimum amount of kinetic energy the vaulter needs to reach this height if air friction is neglected and all the vaulting energy is derived from kinetic energy.

(B) Calculate the speed the vaulter must attain to have the necessary kinetic energy.

7.33 A: (A) $KE = \Delta PE = mg\Delta h$

$$KE = (65kg)(9.8\tfrac{m}{s^2})(5.5m) = 3500J$$

(B) $KE = \tfrac{1}{2}mv^2$

$$v = \sqrt{\frac{2KE}{m}} = \sqrt{\frac{2(3500J)}{65kg}} = 10\tfrac{m}{s}$$

7.34 Q: The work done in accelerating an object along a frictionless horizontal surface is equal to the change in the object's

(1) momentum

(2) velocity

(3) potential energy

(4) kinetic energy

7.34 A: (4) Due to the Work-Energy Theorem.

7.35 Q: A car, initially traveling at 30 meters per second, slows uniformly as it skids to a stop after the brakes are applied. Sketch a graph showing the relationship between the kinetic energy of the car as it is being brought to a stop and the work done by friction in stopping the car.

7.35 A: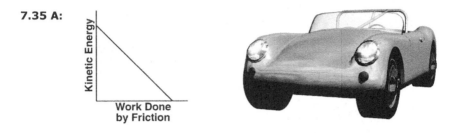

7.36 Q: A 2-kilogram block sliding down a ramp from a height of 3 meters above the ground reaches the ground with a kinetic energy of 50 joules. The total work done by friction on the block as it slides down the ramp is approximately

(1) 6 J

(2) 9 J

(3) 18 J

(4) 44 J

7.36 A: (2) The box has gravitational potential energy at the top of the ramp, which is converted to kinetic energy as it slides down the ramp. Any gravitational potential energy not converted to kinetic energy must be the work done by friction on the block, converted to internal energy (heat) of the system.

$$PE_{top} = KE_{bottom} + W_{friction}$$

$$W_{friction} = PE_{top} - KE_{bottom} = mgh - KE_{bottom}$$

$$W_{friction} = (2kg)(9.8 \, ^m/_{s^2})(3m) - 50J = 9J$$

7.37 Q: Four objects travel down an inclined plane from the same height without slipping. Which will reach the bottom of the incline first?

(1) a baseball rolling down the incline

(2) an unopened soda can rolling down the incline

(3) a physics book sliding down the incline (without friction)

(4) an empty soup can rolling down the incline

7.37 A: (3) In all cases, the objects convert their gravitational potential energy into kinetic energy. In the case of the rolling objects, however, some of that kinetic energy is rotational kinetic energy. Since the physics book cannot rotate, all of its gravitational potential energy becomes translational kinetic energy, therefore it must have the highest translational velocity.

7.38 Q: As a box is pushed 30 meters across a horizontal floor by a constant horizontal force of 25 newtons, the kinetic energy of the box increases by 300 joules. How much total internal energy is produced during this process?

(1) 150 J

(2) 250 J

(3) 450 J

(4) 750 J

7.38 A: (3). The work done on the box can be found from:

$$W = F\Delta r$$

$$W = (25N)(30m) = 750J$$

From the work energy theorem, you know that the total energy of the box must increase by 750 joules. If the kinetic energy of the box increases by 300 joules, where did the other 450 joules of energy go? It must have been transformed into internal energy!

Chapter 8: Fluids

*"What is harder than rock,
or softer than water?
Yet soft water hollows out
hard rock.*

Persevere."

— Ovid

Objectives

1. Calculate the density of an object.
2. Determine whether an object will float given its average density.
3. Calculate the forces on a submerged or partially submerged object using Archimedes' Principle
4. Calculate pressure as the force a system exerts over an area.
5. Explain the operation of a hydraulic system as a function of equal pressure throughout a fluid.
6. Apply the continuity equation to fluids in motion.
7. Apply Bernoulli's Principle to fluids in motion.

If you're going to take a whole chapter to study fluids, it would make sense to start with what fluids are. A **fluid** is matter that flows under pressure, which includes liquids, gases, and even plasmas. Water is a fluid, air is a fluid, the sun is a fluid, even molasses are a fluid. **Fluid Mechanics** is the study of fluids, ranging from fluids at rest, to fluids in motion, to forces applied to and exerted by fluids. You could start your study of fluids in a variety of places, but one of the simplest examples of fluid behavior comes from investigating objects that float and objects that sink. To understand this behavior, why not begin with density?

Density

Density is defined as the ratio of an object's mass to the volume it occupies, and is frequently given the symbol rho (ρ) in physics.

$$\rho = \frac{m}{V}$$

Less dense fluids will float on top of more dense fluids, and less dense solids will float on top of more dense fluids (keeping in mind you must look at the average density of the entire solid object).

8.01 Q: A single kilogram of water fills a cube of length 0.1m. What is the density of water?

8.01 A: $\rho = \dfrac{m}{V}$

$$\rho = \frac{1kg}{(0.1m)(0.1m)(0.1m)} = 1000\,^{kg}\!/_{m^3}$$

8.02 Q: Gold has a density of 19,320 kg/m³. How much volume does a single kilogram of gold occupy?

8.02 A: $\rho = \dfrac{m}{V}$

$$V = \frac{m}{\rho} = \frac{1kg}{19320\,^{kg}\!/_{m^3}} = 5.18 \times 10^{-5}\,m^3$$

8.03 Q: Fresh water has a density of 1000 kg/m³. Which of the following materials will float on water?

(1) Ice (ρ=917 kg/m³)

(2) Magnesium (ρ=1740 kg/m³)

(3) Cork (ρ=250 kg/m³)

(4) Glycerol (ρ=1261 kg/m³)

8.03 A: (1) and (3). Both ice and cork will float on water because they have an average density less than that of water.

8.04 Q: Based on the image below, what can you say about the average density of the man and inner tube compared to the density of the water?

(1) The average density of the man and inner tube is greater than that of the water.

(2) The average density of the man and inner tube is less than that of the water.

(3) The average density of the man and inner tube is equal to that of the water.

8.04 A: (2) The average density of a solid must be less than that of any fluid it is floating in.

Buoyancy

As you can imagine, there is definitely more to whether an object floats or not than just average density. For example, why do some objects float higher in the water than others? And why is it easier to lift objects underwater than in the air? To answer these questions, you'll need to understand the concept of **buoyancy**, a force which is exerted by a fluid on an object, opposing the object's weight.

It is rumored that the Greek philosopher and scientist Archimedes, around 250 B.C., was asked by King Hiero II to help with a problem. King Hiero II had ordered a fancy golden crown from a goldsmith. However, the king was concerned that the goldsmith may have taken his money and mixed some silver in with the crown instead of crafting the crown out of pure gold. He asked Archimedes if there was a way to determine if the crown was pure gold.

Archimedes puzzled over the problem for some time, coming up with the solution while he was in the bath tub one evening. When Archimedes submerged himself in the tub, he noticed that the amount of water that spilled over the rim of the tub was equal to the volume of water he displaced.

Using this method, he could place the crown in a bowl full of water. The amount of water that spilled over could be measured and used to tell the volume of the crown. By then dividing the mass of the crown by the volume, he could obtain the density of the crown, and compare the density to that of gold, determining if the crown was pure gold. According to legend, he was so excited he popped out of the tub and ran through the streets naked yelling "Eureka! Eureka!" (Greek for "I found it! I found it!")

True story or not, this amusing tale illustrates Archimedes' development of a key principle of buoyancy: the buoyant force (F_B) on an object is equal to the density of the fluid, multiplied by the volume of the fluid displaced (which is also equal to the volume of the submerged portion of the object), multiplied by the gravitational field strength. This is known as **Archimedes' Principle**.

$$F_B = \rho_{fluid} V g$$

Archimedes' Principle explains why boats made of steel can float. Although the steel of the boat itself is more dense than water, the average density of the entire boat (including the air in the interior of the boat) is less than that of water. Put another way, the boat floats because the weight of the volume of water displaced by the boat is greater than the weight of the boat itself.

This principle also accounts for the ability of submarines to control their depth. Submarines use pumps to move water into and out of chambers in their interior, effectively controlling the average density of the submarine. If the submarine wants to rise, it pumps water out, reducing its average density. If it wants to submerge, it pumps water in, increasing its average density.

8.05 Q: What is the buoyant force on a 0.3 m³ box which is fully submerged in freshwater (density=1000 kg/m³)?

8.05 A: $F_B = \rho_{fluid} V g = (1000 \, \sfrac{kg}{m^3})(0.3m^3)(9.8 \, \sfrac{m}{s^2}) = 2940N$

8.06 Q: A steel cable holds a 120-kg shark tank 3 meters below the surface of saltwater. If the volume of water displaced by the shark tank is 0.1 m³, what is the tension in the cable? Assume the density of saltwater is 1025 kg/m³.

8.06 A: First, draw a free body diagram (FBD) of the situation, realizing that you have the force of gravity (mg) pulling down, the buoyant force upward, and the force of tension in the cable upward.

Because the shark tank is at equilibrium under the water, the net force on it must be zero, therefore the upward forces must balance the downward forces. You can write this using Newton's 2nd Law in the y-direction as:

$$F_{NET_y} = F_T + F_B - mg = 0$$

Finally, you can use this equation to solve for the force of tension in the cable.

$$F_T = mg - F_B$$
$$F_T = mg - \rho_{fluid} V g$$
$$F_T = (120kg)(9.8 \, ^m/_{s^2}) - (1025 \, ^{kg}/_{m^3})(0.1m^3)(9.8 \, ^m/_{s^2})$$
$$F_T = 172N$$

8.07 Q: A rectangular boat made out of concrete with a mass of 3000 kg floats on a freshwater lake (ρ=1000 kg/m³). If the bottom area of the boat is 6 m², how much of the boat is submerged?

8.07 A: Because the boat is floating on the lake, the magnitude of the buoyant force must be equal to the magnitude of the weight of the boat. (F_B=mg).

Since the boat is rectangular, you can write its volume (V) as its bottom area (A=6 m²) multiplied by the depth submerged (d).

$$F_B = mg$$
$$\rho_{fluid} V g = mg$$
$$\rho_{fluid}(Ad)g = mg$$
$$d = \frac{m}{\rho_{fluid} A} = \frac{3000kg}{(1000 \, ^{kg}/_{m^3})(6m^2)} = 0.5m$$

Pressure

Everyone's been under pressure at one time or another, or in certain circumstances have really "felt the pressure." From a scientific perspective, however, pressure has a very specific definition, and its exploration leads to some very important applications.

In physics, pressure is the effect of a force acting upon a surface. Mathematically, it is a scalar quantity calculated as the force applied per unit area, where the force applied is always perpendicular to the surface. The SI unit of pressure, a Pascal (Pa), is equivalent to a N/m².

$$P = \frac{F}{A}$$

All states of matter can exert pressure. When you walk across an ice-covered lake, you are applying a pressure to the ice equal to the force of gravity on your body (your weight) divided by the area over which you're contacting the ice. This is why it is important to spread your weight out when traversing fragile surfaces. Your odds of breaking through the ice go up tremendously if you walk across the ice in high heels, as the small area contacting the ice leads to a high pressure. This is also the reason snow shoes have such a large area. They are designed to reduce the pressure applied to the top crust of snow so that you can walk more easily without sinking into snow drifts.

Fluids, also, can exert pressure. All fluids exert outward pressure in all directions on the sides of any container holding the fluid. Even the Earth's atmosphere exerts pressure, which you are experiencing right now. The pressures inside and outside your body are so well balanced, however, that you rarely notice the 101,325 Pascals due to the atmosphere (approximately 10N/cm²). If you ride in an airplane and change altitude (and therefore pressure) quickly, you may have experienced a "popping" sensation in your ears — this is due to the pressure inside your ear balancing the pressure outside your ear in a transfer of air through small tubes that connect your inner ear to your throat.

8.08 Q: Air pressure is approximately 100,000 Pascals. What force is exerted on this book when it is sitting flat on a desk? The area of the book's cover is 0.035 m².

8.08 A: $P = \frac{F}{A}$

$F = PA = (100,000 Pa)(0.035 m^2) = 3500 N$

8.09 Q: A fisherman with a mass of 75kg falls asleep on his four-legged chair of mass 5 kg. If each leg of the chair has a surface area of 2.5×10⁻⁴ m² in contact with the ground, what is the average pressure exerted by the fisherman and chair on the ground?

8.09 A: The force applied is the force of gravity, therefore we can write:

$$P = \frac{F}{A} = \frac{mg}{A} = \frac{(75kg + 5kg)(9.8\,^m/_{s^2})}{4(2.5 \times 10^{-4}m^2)} = 784,000\,Pa$$

8.10 Q: A scale which reads 0 in the vacuum of space is placed on the surface of planet Physica. On the planet's surface, the scale indicates a force of 10,000 Newtons. Calculate the surface area of the scale, given that atmospheric pressure on the surface of Physica is 80,000 Pascals.

8.10 A:
$$P = \frac{F}{A}$$
$$A = \frac{F}{P} = \frac{10,000N}{80,000Pa} = 0.125m^2$$

8.11 Q: Rank the following from highest pressure to lowest pressure upon the ground:

(A) The atmosphere at sea level

(B) A 7000-kg elephant with total area 0.5 m² in contact with the ground

(C) A 65-kg lady in high heels with total area 0.005 m² in contacting with the ground

(D) A 1600-kg car with a total tire contact area of 0.2 m²

8.11 A: (1) B. The elephant (137,000 Pa)

(2) C. The lady in high heels (127,000 Pa)

(3) A. The atmosphere (100,000 Pa)

(4) D. The car (78,400 Pa)

The pressure that a fluid exerts on an object submerged in that fluid can be calculated almost as simply. If the object is submersed to a depth (h), the pressure is found by multiplying the density of the fluid by the depth submerged, all multiplied by the acceleration due to gravity.

$$P_{gauge} = \rho g h$$

This is known as the gauge pressure, because this is the reading you would observe on a pressure gauge. If there is also atmosphere above the fluid, such as the situation here on earth, you can determine the absolute pressure, or total pressure, by adding in the atmospheric pressure (P_0), which is equal to approximately 100,000 Pascals.

$$P_{absolute} = P_0 + P_{gauge} = P_0 + \rho g h$$

8.12 Q: Samantha spots buried treasure while scuba diving on her Caribbean vacation. If she must descend to a depth of 40 meters to examine the pressure, what gauge pressure will she read on her scuba equipment? The density of sea water is 1025 kg/m³.

8.12 A: $P_{gauge} = \rho g h$

$P_{gauge} = (1025\,{}^{kg}\!/_{m^3})(9.8\,{}^{m}\!/_{s^2})(40m) = 402,000\,Pa$

8.13 Q: What is the absolute pressure exerted on the diver in the previous problem by the water and atmosphere?

8.13 A:

$P_{absolute} = P_0 + P_{gauge}$

$P_{absolute} = 100,000\,Pa + 402,000\,Pa = 502,000\,Pa$

8.14 Q: A diver's pressure gauge reads 250,000 Pascals in fresh water (ρ=1000 kg/m³). How deep is the diver?

8.14 A: $\quad P = \rho g h$

$$h = \frac{P}{\rho g} = \frac{250,000\,Pa}{(1000\,{}^{kg}\!/_{m^3})(9.8\,{}^{m}\!/_{s^2})} = 25.5m$$

Pascal's Principle

When a force is applied to a contained, incompressible fluid, the pressure increases equally in all directions throughout the fluid. This fundamental characteristic of fluids provides the foundation for hydraulic systems found in barbershop chairs, construction equipment, and the brakes in your car.

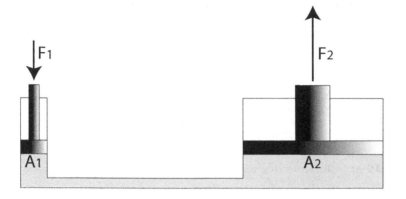

Because the force applied to the contained fluid is distributed throughout the system, you can multiply the applied force through this application of Pascal's Principle in the following manner. Assume you have a closed container filled with an incompressible fluid with two pistons of differing areas, A_1 and A_2. If you apply a force, F_1, to the piston of area A_1, you create a pressure in the fluid which you can call P_1.

$$P_1 = \frac{F_1}{A_1}$$

Similarly, the pressure at the second piston, P_2, must be equal to F_2 divided by the area of the second piston, A_2.

$$P_2 = \frac{F_2}{A_2}$$

Since the pressure is transmitted equally throughout the fluid in all directions according to Pascal's Principle, P_1 must equal P_2.

$$P_1 = P_2 \rightarrow \frac{F_1}{A_1} = \frac{F_2}{A_2}$$

Rearranging to solve for F_2, you find that F_2 is increased by the ratio of the areas A_2 over A_1.

$$F_2 = \frac{A_2}{A_1} F_1$$

Therefore, you have effectively increased the applied force F_1. Of course, the law of conservation of energy cannot be violated, so the work done on the system must balance the work done by the system. In the hydraulic lift diagram shown on the previous page, the distance over which F_1 is applied will be greater than the distance over which F_2 is applied, by the exact same ratio as the force multiplier!

8.15 Q: A barber raises his customer's chair by applying a force of 150N to a hydraulic piston of area 0.01 m². If the chair is attached to a piston of area 0.1 m², how massive a customer can the chair raise? Assume the chair itself has a mass of 5 kg.

8.15 A: To solve this problem, first determine the force applied to the larger piston.

$$F_2 = \frac{A_2}{A_1} F_1$$

$$F_2 = \frac{0.10m^2}{0.01m^2} (150N) = 1500N$$

If the maximum force on the chair is 1500N, you can now determine the maximum mass which can be lifted by recognizing that the force that must be overcome to lift the customer is the force of gravity, therefore the applied force on the customer must equal the force of gravity on the customer.

$$F = mg$$

$$m = \frac{F}{g} = \frac{1500N}{9.8 \, m/_{s^2}} = 153kg$$

If the chair has a mass of 5 kilograms, the maximum mass of a customer in the chair must be 148 kg.

8.16 Q: A hydraulic system is used to lift a 2000-kg vehicle in an auto garage. If the vehicle sits on a piston of area 1 square meter, and a force is applied to a piston of area 0.03 square meters, what is the minimum force that must be applied to lift the vehicle?

    (1) 11,600 N

    (2) 3330 N

    (3) 1180 N

    (4) 120 N

8.16 A: (3) 1180 N

$$P_1 = P_2 \rightarrow \frac{F_1}{A_1} = \frac{F_2}{A_2}$$

$$F_1 = \frac{A_1}{A_2} F_2 = \frac{0.03m^2}{0.5m^2}\left(2000kg \times 9.8\,^m\!/_{s^2}\right)$$

$$F_1 = 1180N$$

Continuity Equation for Fluids

When fluids move through a full pipe, the volume of fluid that enters the pipe must equal the volume of fluid that leaves the pipe, even if the diameter of the pipe changes. This is a restatement of the law of conservation of mass for fluids.

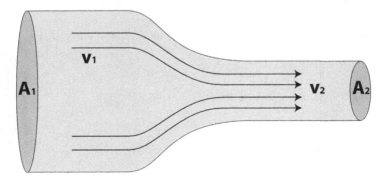

The volume of fluid moving through the pipe at any point can be quantified in terms of the volume flow rate, which is equal to the diameter of the pipe at that point multiplied by the velocity of the fluid. This volume flow rate must be constant throughout the pipe, therefore you can write the equation of continuity for fluids (also known as the fluid continuity equation) as:

$$A_1 v_1 = A_2 v_2$$

This equation says that as the cross-section of the pipe gets smaller, the velocity of the fluid increases, and as the cross-section gets larger, the fluid velocity decreases. You may have applied this yourself in watering the flowers with a garden hose. If you want increase the velocity of the water coming from the end of the hose, you place your thumb over part of the opening of the hose, effectively decreasing the cross-sectional area of the hose's end and increasing the velocity of the exiting water!

8.17 Q: Water runs through a water main of cross-sectional area 0.4 m² with a velocity of 6 m/s. Calculate the velocity of the water in the pipe when the pipe tapers down to a cross-sectional area of 0.3 m².

(1) 4.5 m/s

(2) 6 m/s

(3) 8 m/s

(4) 10.7 m/s

8.17 A: (3) $A_1 v_1 = A_2 v_2$

$$v_2 = \frac{A_1}{A_2} v_1 = \frac{0.4m^2}{0.3m^2} (6\,{}^m\!/_s) = 8\,{}^m\!/_s$$

8.18 Q: Water enters a typical garden hose of diameter 1.6 cm with a velocity of 3 m/s. Calculate the exit velocity of water from the garden hose when a nozzle of diameter 0.5 cm is attached to the end of the hose.

8.18 A: First, find the cross-sectional areas of the entry (A_1) and exit (A_2) sides of the hose.

$$A_1 = \pi r^2 = \pi (0.008m)^2 = 2 \times 10^{-4} m^2$$

$$A_2 = \pi r^2 = \pi (0.0025m)^2 = 1.96 \times 10^{-5} m^2$$

Next, apply the continuity equation for fluids to solve for the water velocity as it exits the hose (v_2).

$$A_1 v_1 = A_2 v_2$$

$$v_2 = \frac{A_1}{A_2} v_1 = \frac{2 \times 10^{-4} m^2}{1.96 \times 10^{-5} m^2} (3\,{}^m\!/_s) = 30.6\,{}^m\!/_s$$

Bernoulli's Principle

Conservation of energy, when applied to fluids in motion, leads to Bernoulli's Principle. **Bernoulli's Principle** states that fluids moving at higher velocities lead to lower pressures, and fluids moving at lower velocities result in higher pressures.

Airplane wings have a larger top surface than a bottom surface to take advantage of this fact. As the air moves across the larger top surface, it must move faster than the air traveling a shorter distance under the bottom surface. This leads to a lower pressure on top of the wing, and a higher pressure underneath the wing, providing some of the lift for the aircraft (note that this isn't the only cause of lift, as Newton's 3rd Law also plays a critical role in understanding the dynamics of flight).

This principle is also used in sailboats, carburetors, gas delivery systems, and even water-powered sump pumps!

Expressing Bernoulli's Principle quantitatively, you can relate the pressure, velocity, and height of a liquid in a tube at various points.

$$P_1 + \tfrac{1}{2}\rho v_1^2 + \rho g y_1 = P_2 + \tfrac{1}{2}\rho v_2^2 + \rho g y_2$$

The pressure at a point in the tube plus half the density of the fluid multiplied by the square of its velocity at that point, added to the gauge pressure of the fluid ($\rho g y$), must be equal at any point in the tube.

8.19 Q: Water sits in a large open jug at a height of 0.2m above the spigot. With what velocity will the water leave the spigot when the spigot is opened?

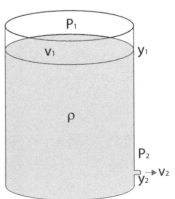

8.19 A: Since the top of the jug and the spigot are both open to atmosphere, the pressures P_1 and P_2 must be equal. Since the jug is much larger than the spigot, you can assume the velocity of the water at the top of the jug is nearly zero. This allows you to simplify Bernoulli's Equation considerably.

$$P_1 + \tfrac{1}{2}\rho v_1^2 + \rho g y_1 = P_2 + \tfrac{1}{2}\rho v_2^2 + \rho g y_2$$

$$\rho g y_1 = \tfrac{1}{2}\rho v_2^2 + \rho g y_2$$

Since the density of the fluid is the same throughout, you can do some algebraic simplification to solve for v_2.

$$g(y_1 - y_2) = \tfrac{1}{2}v_2^2$$

$$v_2 = \sqrt{2g(y_1 - y_2)}$$

This is known as Torricelli's Theorem. Since the difference in height is 0.2m, you can now easily solve for the velocity of the water at the spigot.

$$v_2 = \sqrt{2g(y_1 - y_2)}$$

$$v_2 = \sqrt{2(9.8\,{}^m\!/_{s^2})(0.2m)} = 1.98\,{}^m\!/_s$$

Notice that this is the same result you would obtain if you had solved for the velocity of an object dropped from a height of 0.2 meters using the kinematic equations... this should make sense, as Bernoulli's Equation is really just a restatement of conservation of energy, applied to fluids!

Chapter 9: Thermal Physics

*"Thermodynamics is a funny subject.
The first time you go through it,
you don't understand it at all.*

*The second time you go through it,
you think you understand it,
except for one or two small points.*

*The third time you go through it,
you know you don't understand it,
but by that time you are so used to it,
it doesn't bother you anymore."*

— Arnold Sommerfeld

Objectives

1. Calculate the temperature of an object given its average kinetic energy.
2. Calculate the linear and volumetric expansion of a solid as a function of temperature.
3. Explain heat as the process of transferring energy between systems at different temperatures.
4. Calculate an object's temperature change using its specific heat.
5. Determine the energy required for a material to undergo a phase change.
6. Utilize the ideal gas law to solve for pressure, volume, temperature, and quantity of an ideal gas.
7. Describe the zeroth, first, second, and third laws of thermodynamics.
8. Utilize PV diagrams to describe changes in ideal gas conditions.
9. Analyze adiabatic, isobaric, isochoric, and isothermal processes using both algebraic and graphical methods.

Thermal physics deals with the internal energy of objects due to the motion of the atoms and molecules comprising the objects, as well as the transfer of this energy from object to object, known as heat.

Temperature

The internal energy of an object, known as its thermal energy, is related to the kinetic energy of all the particles comprising the object. The more kinetic energy the constituent particles have, the greater the object's thermal energy.

In solids, the particles comprising the solid are held together tightly, therefore their motion is limited to vibrating back and forth in their given positions. In liquids, the particles can move back and forth across each other, but the object itself has no defined shape. In gases, the particles move throughout the volume available, interacting with each other and the walls of any container holding them. In all cases, the total thermal energy of the object is the sum total of the kinetic energies of its constituent particles.

Instead of just looking at the sum of all the individual particles' kinetic energies, you could examine the average kinetic energy of the particles comprising the object, realizing that the actual kinetic energies of individual particles may vary significantly. The average kinetic energy of the particles is directly related to the temperature of the object by the following equation:

$$K_{avg} = \frac{3}{2} k_B T$$

Examining this equation, the average kinetic energy is given in Joules, k_B is Boltzmann's Constant (1.38×10^{-23} J/K), and the temperature is given in Kelvins, the SI unit of temperature. Note that even though two objects can have the same temperature (and therefore the same average kinetic energy), they may have different internal energies.

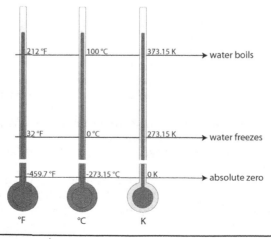

The Kelvin scale is closely related to the Celsius temperature scale, but where the Celsius scale targets the freezing point of water as 0° C, the Kelvin scale utilizes its zero at what is known as absolute zero (the point on a Volume vs. Temperature graph for a gas where the extended curve would hypothetically reach zero volume), considered a theoretical minimum temperature. Therefore, absolute zero is 0 Kelvins, which is equivalent to -273.15° Celsius. To convert from Kelvins to degrees Celsius, just add 273.15 to your temperature reading in degrees Celsius. The freezing point of water, then, is 0°C or 273.15 K, and the boiling point of water is 100°C or 373.15 K. Compare this to the Fahrenheit scale, where water freezes at 32°F, and boils at 212°F!

$$T_{K} = T_{^\circ C} + 273.15$$

$$T_{^\circ F} = \tfrac{9}{5}T_{^\circ C} + 32$$

$$T_{^\circ C} = \tfrac{5}{9}(T_{^\circ F} - 32)$$

9.01 Q: What is the average kinetic energy of the molecules in a steak at a temperature of 345 Kelvins?

(1) 223 J

(2) 4.76×10⁻²¹ J

(3) 7.14×10⁻²¹ J

(4) 517 J

9.01 A: (3) $K_{avg} = \dfrac{3}{2}k_{B}T = \dfrac{3}{2}(1.38 \times 10^{-23} \, {}^{J}\!/_{K})(345K) = 7.14 \times 10^{-21} J$

9.02 Q: Normal canine body temperature is 101.5°F. What is normal canine body temperature in degrees Celsius? In Kelvins?

9.02 A: $T_{^\circ C} = \tfrac{5}{9}(T_{^\circ F} - 32) = \tfrac{5}{9}(101.5 - 32) = 38.6°C$

$T_{K} = T_{^\circ C} + 273.15 = 38.6 + 273.15 = 311.75K$

9.03 Q: The average temperature of space is estimated as roughly -270°C. What is the average kinetic energy of the particles in space?

9.03 A: First, convert the temperature into Kelvins.

$$T_K = T_{°C} + 273.15 = -270.4 + 273.15 = 2.75K$$

Next, solve for the average kinetic energy of the particles.

$$K_{avg} = \frac{3}{2}k_B T = \frac{3}{2}(1.38 \times 10^{-23}\,{}^{J}\!/\!_K)(2.75K) = 3.80 \times 10^{-23}\,J$$

9.04 Q: Given that the average kinetic energy of the particles comprising our sun is 1.2×10⁻¹⁹ J, find the temperature of the sun.

9.04 A: $K_{avg} = \frac{3}{2}k_B T$

$$T = \frac{2K_{avg}}{3k_B} = \frac{2(1.2 \times 10^{-19}\,J)}{3(1.38 \times 10^{-23}\,{}^{J}\!/\!_K)} = 5800K$$

9.05 Q: Which graph best represents the relationship between the average kinetic energy (K_avg) of the random motion of the molecules of an ideal gas and its absolute temperature (T)?

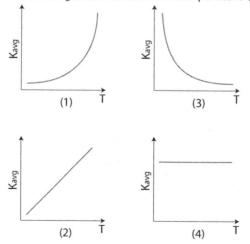

9.05 A: (2) since K_avg is a linear function of absolute temperature T.

9.06 Q: While orbiting Earth, the space shuttle has recorded temperatures ranging from 398K to 118K. These temperatures correspond to Celsius temperatures ranging from

(1) 125°C to -391°C

(2) 125°C to -155°C

(3) 671°C to 391°C

(4) 671°C to 155°C

9.06 A: (2) 125°C to -155°C

9.07 Q: The temperature at which no thermal energy can be transferred from one object to another is

(1) -273 K

(2) 0 K

(3) 0°C

(4) 273°C

9.07 A: (2) 0 K is absolute zero.

Thermal Expansion

When objects are heated, they tend to expand, and when they are cooled, they tend to contract. You can use this to open glass jars with tight metal lids by running the lids under hot water. The temperature increase in the lid expands the metal lid and the glass jar, but because most metals expand more quickly than glass, the lid becomes looser, making it easier to open the jar.

This occurs because at higher temperatures, objects have higher kinetic energies, so their particles vibrate more. At these higher levels of vibration, the particles aren't bound to each other as tightly, so the object expands.

The amount an object expands can be calculated for both one-dimensional (linear) and three-dimensional (volumetric) expansion. The amount a material expands is characterized by the material's coefficient of expansion. In calculating a material's one-dimensional expansion, you can use the linear expansion formula, and the material's linear coefficient of expansion (α).

$$\Delta l = \alpha l_0 \Delta T$$

When calculating a material's three-dimensional expansion, you'll use the volumetric expansion formula, and the material's volumetric coefficient of expansion (β). Note that in most cases, the volumetric coefficient of expansion is roughly three times the linear coefficient of expansion.

$$\Delta V = \beta V_0 \Delta T$$

The change in temperature in the expansion equations can be given in either degrees Celsius or Kelvins.

A sample table showing coefficients of thermal expansion for selected materials is given below.

Approximate Coefficients of Thermal Expansion at 20°C		
Material	α (10⁻⁶/°C)	β (10⁻⁶/°C)
Aluminum	23	69
Concrete	12	36
Diamond	1	3
Glass	9	27
Stainless Steel	17	51
Water*	69	207

Water actually expands when it freezes, so calculations near the freezing point of water require a more detailed analysis than is provided here.

9.08 Q: A concrete railroad tie has a length 2.45 meters on a hot, sunny, 35°C day. What is the length of the railroad tie in the winter when the temperature dips to -25°C?

9.08 A: First find the change in the tie's length.

$$\Delta l = \alpha l_0 \Delta T = (12 \times 10^{-6}\,/_{°C})(2.45m)(-60°C)$$
$$\Delta l = -0.0018m$$

Then, you can find the railroad tie's final length using the tie's initial length and its change in length.

$$\Delta l = l - l_0 \rightarrow l = l_0 + \Delta l$$
$$l = 2.45m + (-0.0018m) = 2.448m$$

9.09 Q: An aluminum rod has a length of exactly one meter at 300K. How much longer is it when placed in a 400°C oven?

9.09 A: Since the temperatures are given in two different sets of units, you first need to find the total temperature shift in consistent units (for example, Kelvins).

$$T_K = T_{°C} + 273.15 = 400 + 273.15 = 673.15K$$

The shift in temperature, therefore, must be 673.15K-300K, or 373.15K. Next, you can use the equation for linear expansion to find the shift in the rod's length.

$$\Delta l = \alpha l_0 \Delta T = (23 \times 10^{-6} /_K)(1m)(373.15K)$$
$$\Delta l = 0.0086m$$

9.10 Q: A glass of water with volume 1 liter is completely filled at 5°C. How much water will spill out of the glass when the temperature is raised to 85°C?

9.10 A: In this situation, both the glass and the water within will expand as the temperature rises. You can treat both the glass and the water as a volume expansion. Start by finding the expansion of the water.

$$\Delta V = \beta V_0 \Delta T = (207 \times 10^{-6} /_{°C})(1L)(80°C) = 0.017L$$

In a similar fashion, you can find the expansion of the glass.

$$\Delta V = \beta V_0 \Delta T = (27 \times 10^{-6} /_{°C})(1L)(80°C) = 0.002L$$

The amount of water spilling out is equal to the difference between the water's expansion and the glass's expansion, or 0.015 liters.

Heat

Heat is the transfer of thermal energy from one object to another object due to a difference in temperature. Heat always flows from warmer objects to cooler objects. The symbol for heat in physics is Q, with positive values of Q representing heat flowing into an object, and negative values of Q representing heat flowing out of an object.

When heat flows into or out of an object, the amount of temperature change depends on the material. The amount of heat required to change one kilogram of a material by one degree Celsius (or one Kelvin) is known as the material's specific heat (or specific heat capacity), represented by the symbol C.

Specific Heats of Selected Materials	
Material	**C (J/kg·K)**
Aluminum	897
Concrete	850
Diamond	509
Glass	840
Helium	5193
Water	4181

The relationship between heat and temperature is quantified by the following equation, where Q is the heat transferred, m is the mass of the object, C is the specific heat, and ΔT is the change in temperature (in degrees Celsius or Kelvins).

$$Q = mC\Delta T$$

9.11 Q: A half-carot diamond (0.0001 kg) absorbs five Joules of heat. How much does the temperature of the diamond increase?

9.11 A: $Q = mC\Delta T \rightarrow \Delta T = \dfrac{Q}{mC}$

$$\Delta T = \frac{5J}{(0.0001kg)(509\,{}^{J}\!/_{kg \bullet K})} = 98K$$

9.12 Q: A three-kilogram aluminum pot is filled with five kilograms of water. How much heat is absorbed by the pot and water when both are heated from 25°C to 95°C?

9.12 A: Using the table of specific heats, you can find the heat added to each item separately, and then combine them to get the total heat added.

$$Q_{Al} = m_{Al}C_{Al}\Delta T = (3kg)(897\,{}^{J}\!/_{kg \bullet K})(70K) = 1.88 \times 10^{5}J$$

$$Q_{H_2O} = m_{H_2O}C_{H_2O}\Delta T = (5kg)(4181\,{}^{J}\!/_{kg \bullet K})(70K) = 1.46 \times 10^{6}J$$

The total heat absorbed, therefore, must be 1.65×10⁶ Joules.

9.13 Q: Two solid metal blocks are placed in an insulated container. If there is a net flow of heat between the blocks, they must have different

(1) initial temperatures
(2) melting points
(3) specific heats
(4) heats of fusion

9.13 A: (1) since heat flows from warmer objects to cooler objects.

Heat can be transferred from one object to another by three different methods: conduction, convection, and radiation. **Conduction** is the transfer of heat along an object due to the particles comprising the object colliding. When you stick an iron rod in a fire, the end in the fire warms up, but over time, the particles comprising the iron rod near the fire move more quickly, colliding with other particles in the iron speeding them up, and so on, and so on, resulting in heat transfer down the length of the iron rod until the end you're holding far away from the fire becomes very hot!

Convection, on the other hand, is a result of the energetic (heated) particles moving from one place to another. A great example of this is a convection oven. In a convection oven, air molecules are heated near the burner or electrical element, and then circulated throughout the oven, transferring the heat throughout the entire oven's volume.

Radiation is the transfer of heat through electromagnetic waves. Think of a campfire or fireplace on a cold evening. When you want to warm up, you place your hands up in front of you, allowing your hands to absorb the maximum amount of electromagnetic waves (mostly infrared) coming from the fire, making you nice and toasty!

Looking more closely at conduction specifically, the rate of heat transfer (H), as measured in Joules per second, or Watts, depends on the magnitude of the temperature difference across the object (ΔT), the cross-sectional area of the object (A), the length of the object (L) and the thermal conductivity of the material (k). Thermal conductivities are typically provided to you in a problem, or you can look them up in a table of thermal conductivities.

$$H = \frac{kA\Delta T}{L}$$

Thermal Conductivities of Selected Materials	
Material	**k (J/s·m·K)**
Aluminum	237
Concrete	1
Copper	386
Glass	0.9
Stainless Steel	16.5
Water	0.6

9.14 Q: Find the rate of heat transfer through a 5 mm thick glass window with a cross-sectional area of 0.4 m² if the inside temperature is 300K and the outside temperature is 250K.

9.14 A: $H = \dfrac{kA\Delta T}{L} = \dfrac{(0.9\,{}^{J}\!/_{s \bullet m \bullet K})(0.4m^2)(300\,K - 250\,K)}{0.005m} = 3600W$

9.15 Q: One end of a 1.5-meter-long stainless steel rod is placed in an 850K fire. The cross-sectional radius of the rod is 1 cm, and the cool end of the rod is at 300K. Calculate the rate of heat transfer through the rod.

9.15 A: To solve this problem, you must first find the cross-sectional area of the rod.

$$A = \pi r^2 = \pi(.01m)^2 = 3.14 \times 10^{-4} m^2$$

Next, calculate the heat transfer through the rod.

$$H = \frac{kA\Delta T}{L}$$

$$H = \frac{(16.5\,{}^{J}\!/_{s \bullet m \bullet K})(3.14 \times 10^{-4} m^2)(850\,K - 300\,K)}{1.5m} = 1.9W$$

Phase Changes

As you know, matter can exist in different states. These states include solids, liquids, gases, and plasmas. You're probably familiar with solids, liquids, and gases already. Plasmas are energetic gases that have been ionized so that they can conduct electricity (examples include stars, lightning, neon signs, etc.)

When matter changes state, its internal energy changes, so the kinetic energy of its constituent particles changes. As it is changing from one state to another, the change in energy is reflected in the bonds between the particles, and therefore the temperature of the object doesn't change. Once the state change is complete, however, changes in energy are again observed in the form of changes in temperature.

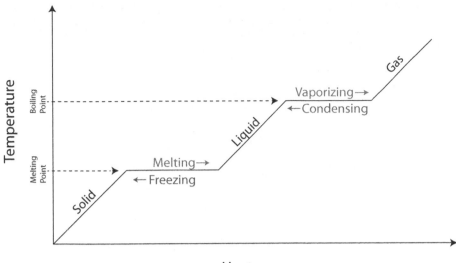

The energy required to change a specific material's state is known as the material's latent heat of transformation (L). When an object transitions from the solid to liquid phase, you use the latent heat of fusion (L_f). When an object transitions from the liquid to the gaseous phase, you use the latent heat of vaporization (L_v). You can calculate the energy required for a material to change phases using the following formula, where Q is the heat added, m is the object's mass, and L is the material's specific latent heat of transformation.

$$Q = mL$$

Latent Heats of Selected Materials				
Material	**L_f (kJ/kg)**	**Melting Point (°C)**	**L_v (kJ/kg)**	**Boiling Point (°C)**
Aluminum	399	659	10,500	2327
Helium	N/A	N/A	21	-269
Hydrogen	58	-259	455	-253
Lead	25	327	871	1750
Water	334	0	2260	100

9.16 Q: The graph below represents a cooling curve for 10 kilograms of a substance as it cools from a vapor at 160°C to a solid at 20°C. Energy is removed from the sample at a constant rate.

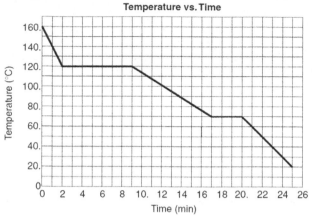

Temperature vs. Time

While the substance is cooling during the liquid phase, the average kinetic energy of the molecules of the substance

(1) decreases

(2) increases

(3) remains the same

9.16 A: (1) decreases since the temperature decreases, and average kinetic energy is related to temperature.

9.17 Q: Based on the graph of the previous problem, the melting point of the substance is

(1) 0°C

(2) 70°C

(3) 100°C

(4) 120°C

9.17 A: (2) 70°C.

9.18 Q: How much heat must be added to a 10 kg lead bar to change the bar from a solid to a liquid at 327°C?

9.18 A: $Q_f = mL_f = (10kg)(25000 \, ^{J}\!/_{kg}) = 250,000 J$

9.19 Q: How much heat must be added to 1 kg of water to change it from a 50°C to 100°C steam at standard pressure?

9.19 A: To solve this problem, you must find both the amount of heat required to change the temperature of the water, as well as the amount of heat required to change the state of the water.

$$Q = mC\Delta T + mL_v = m(C\Delta T + L_v)$$
$$Q = 1kg(4181\,{}^{J}\!/_{kg \bullet K} \times 50°C + 2260000\,{}^{J}\!/_{kg}) = 2.47 \times 10^6 J$$

9.20 Q: The graph below shows temperature vs. time for one kilogram of an unknown material as heat is added at a constant rate.

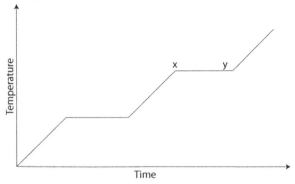

During interval *xy*, the material experiences
(1) a decrease in internal energy and a phase change
(2) an increase in internal energy and a phase change
(3) no change in internal energy and a phase change
(4) no change in internal energy and no phase change

9.20 A: (2) an increase in internal energy and a phase change.

Ideal Gas Law

In studying the behavior of gases in confined spaces, it is useful to limit ourselves to the study of ideal gases. **Ideal gases** are theoretical models of real gases, which utilize a number of basic assumptions to simplify their study. These assumptions include treating the gas as being comprised of many particles which move randomly in a container. The particles are, on average, far apart from one another, and they do not exert forces upon one another unless they come in contact in an elastic collision.

Under normal conditions such as standard temperature and pressure, most gases behave in a manner quite similar to an ideal gas. Heavy gases as well as gases at very low temperatures or very high pressures are not as well modeled by an ideal gas.

The **Ideal Gas Law** relates the pressure, volume, number of particles, and temperature of an ideal gas in a single equation, and can be written in a number of different ways.

$$PV = nRT = Nk_BT$$

In this equation, P is the pressure of the gas (in Pascals), V is the volume of the gas (in cubic meters), n is the number of moles of gas, N is the number of molecules of gas, R is the universal gas constant equal to 8.31 J/mol·K (which is also 0.08206 L·atm/mol·K), k_B is Boltzmann's Constant (1.38×10^{-23} J/K), and T is the temperature, in Kelvins. To convert from molecules to moles, you can use Avogadro's Number ($N_0 = 6.02 \times 10^{23}$ molecules/mole):

$$n = \frac{N}{N_0}$$

Note that a Pascal multiplied by a cubic meter is a newton-meter, or Joule. As well, Boltzmann's constant is the ideal gas law constant divided by Avogadro's number.

9.21 Q: How many moles of an ideal gas are equivalent to 3.01×10^{24} molecules?

9.21 A: $n = \dfrac{N}{N_0} = \dfrac{3.01 \times 10^{24} \, molecules}{6.02 \times 10^{23} \, molecules/mole} = 5 \, moles$

9.22 Q: Find the number of molecules in 0.4 moles of an ideal gas.

9.22 A: $n = \dfrac{N}{N_0} \rightarrow N = nN_0 \rightarrow$

$N = (0.4 \, moles)(6.02 \times 10^{23} \, molecules/mole) = 2.41 \times 10^{23} \, molecules$

9.23 Q: How many moles of gas are present in a 0.3 m³ bottle of carbon dioxide held at a temperature of 320K and a pressure of 1×10^6 Pascals?

9.23 A: $PV = nRT \rightarrow n = \dfrac{PV}{RT} = \dfrac{(1 \times 10^6 \, Pa)(0.3m^3)}{(8.31 \, J/mol \cdot K)(320K)} = 113 \, moles$

9.24 Q: A cubic meter of carbon dioxide gas at room temperature (300K) and atmospheric pressure (101,325 Pa) is compressed into a volume of 0.1 m³ and held at a temperature of 260K. What is the pressure of the compressed carbon dioxide?

9.24 A: Since the number of moles of gas is constant, you can simplify the ideal gas equation into the combined gas law by setting the initial pressure, volume, and temperature relationship equal to the final pressure, volume, and temperature relationship.

$$\frac{P_1 V_1}{T_1} = nR = \frac{P_2 V_2}{T_2}$$

Since you know all the quantities in this equation except for the final pressure, you can solve for the final pressure directly.

$$P_2 = \frac{P_1 V_1 T_2}{T_1 V_2} = \frac{(101,325 Pa)(1m^3)(260K)}{(300K)(0.1m^3)} = 878,000 Pa$$

9.25 Q: One mole of helium gas is placed inside a balloon. What is the pressure inside the balloon when the balloon rises to a point in the atmosphere where the temperature is -12°C and the volume of the balloon is 0.25 cubic meters?

9.25 A: First you must convert the temperature from degrees Celsius to Kelvins.

$$T_K = T_{°C} + 273.15 = -12°C + 273.15 = 261.15K$$

Next, you can use the ideal gas law to solve for the pressure inside the balloon.

$$PV = nRT \rightarrow P = \frac{nRT}{V} \rightarrow$$

$$P = \frac{(1mole)(8.31\,{}^J\!/_{mol\bullet K})(261.15K)}{(.25m^3)} = 8680 Pa$$

It's also quite straightforward to find the total internal energy of an ideal gas. Recall that the average kinetic energy of the particles of an ideal gas are described by the formula:

$$K_{avg} = \frac{3}{2}k_B T$$

The total internal energy of an ideal gas can be found by multiplying the average kinetic energy of the gas's particles by the number of particles (N) in the gas. Therefore, the internal energy of the gas can be calculated using:

$$U = N \times K_{av} \xrightarrow[K_{av}=\frac{3}{2}k_B T]{N=nN_0} U = \frac{3}{2}nN_0 k_B T \xrightarrow{N_0 k_B = R}$$

$$U = \frac{3}{2}nRT$$

9.26 Q: Find the internal energy of 5 moles of oxygen at a temperature of 300K.

9.26 A: $U = \frac{3}{2}nRT = \frac{3}{2}(5moles)(8.31\,{}^{J}\!/_{mol\bullet K})(300K) = 18.7kJ$

9.27 Q: What is the temperature of 20 moles of argon with a total internal energy of 100 kJ?

9.27 A: $U = \frac{3}{2}nRT \rightarrow T = \dfrac{2U}{3nR} \longrightarrow$

$$T = \frac{2(100,000J)}{3(20moles)(8.31\,{}^{J}\!/_{mol\bullet K})} = 401K$$

Thermodynamics

Thermodynamics, which began as an effort to increase the efficiency of steam engines in the early 1800s, can be thought of as the study of the relationship between heat transferred to or from an object, and the work done on or by an object. Both heat and work deal with the transfer of energy, but heat involves energy transfer due to a temperature difference.

The zeroth law of thermodynamics (don't blame me, I didn't name it!) states that if object A is in thermal equilibrium with object B, and object B is in thermal equilibrium with object C, then objects A and C must be in thermal equilibrium with each other. This law is so intuitive it almost doesn't need stating, but in defining proofs of the 1st and 2nd laws of thermodynamics, scientists realized they needed this law specifically stated to complete their proofs.

The first law of thermodynamics is really a restatement of the law of conservation of energy. Specifically, it states that the change in the internal energy of a closed system is equal to the heat added to the system plus the work done on the system, and is written as:

$$\Delta U = Q + W$$

In this equation it is important to note the sign conventions, where a positive value for heat, Q, represents heat added to the system, and a positive value for work, W, indicates work done on the gas. If energy were being pulled from the system, as in heat taken from the system or work done by the system, those quantities would be negative.

In most cases, you'll utilize the first law of thermodynamics to analyze the behavior of ideal gases, which can be streamlined by analyzing the definition of work on a gas.

$$W = F\Delta r \xrightarrow{F=PA} W = PA\Delta r \xrightarrow{\Delta V = -A\Delta r} W = -P\Delta V$$

If work is force multiplied by displacement, and pressure is force over area, force can be replaced with pressure multiplied by area. The area multiplied by the displacement gives you the change in volume of the gas. Due to the sign convention that work done on the gas is positive (corresponding to a decrease in volume), you can write work as W=-PΔV.

9.28 Q: Five thousand joules of heat is added to a closed system, which then does 3000 joules of work. What is the net change in the internal energy of the system?

9.28 A: Keeping in mind the positive/negative sign convention:
$$\Delta U = Q + W = 5000J - 3000J = 2000J$$

9.29 Q: A liquid is changed to a gas at atmospheric pressure (101,325 Pa). The volume of the liquid was 5×10^{-6} m³. The volume of gas is 5×10^{-3} m³. How much work was done in the process?

9.29 A: $W = -P\Delta V = -P(V_f - V_i) \rightarrow$
$$W = -(101,325\,Pa)(5\times10^{-3}\,m^3 - 5\times10^{-6}\,m^3) = -506J$$

Pressure-Volume Diagrams (PV diagrams) are useful tools for visualizing the thermodynamic processes of gases. These diagrams show pressure on the y-axis, and volume on the x-axis, and are used to describe the changes undergone by a set amount of gas. Because the amount of gas remains constant, a PV diagram not only tells you pressure and volume, but can also be used to determine the temperature of a gas when combined with the ideal gas law. A Sample PV diagram is shown at right, showing two states of the gas, state A and state B.

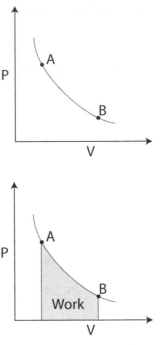

In transitioning from state A to state B, the volume of the gas increases, while the pressure of the gas decreases. In transitioning from state B to state A, the volume of the gas decreases, while the pressure increases. Because the work done on the gas is given by $W=-P\Delta V$, you can find the work done on the gas graphically from the PV diagram by taking the area under the curve. Because of the positive/negative sign convention, as the volume of gas expands the gas does work (W is negative), and as the gas compresses, work is done on the gas (W is positive).

9.30 Q: Using the PV diagram below, find the amount of work required to transition from state A to B, and then the amount of work required to transition from state B to state C.

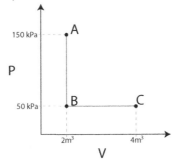

9.30 A: The amount of work in moving from state A to B is equal to the area under the graph for that transition. Since there is no area under the straight line, no work was done. The work in moving from state B to state C can be found by taking the area under the line in the PV diagram.

$$W = -P\Delta V = -P(V_f - V_i) \rightarrow$$

$$W = -(50000\,Pa)(4m^3 - 2m^3) = -1 \times 10^5\,J$$

Note that the work is negative, indicating the gas did work, which correlates with the gas expanding.

In exploring ideal gas state changes, there are a number of state changes in which one of the characteristics of the gas or process remain constant, and are illustrated on the PV diagram below.

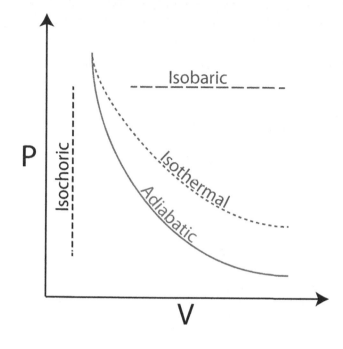

The types of processes include:

- **Adiabatic** — Heat (Q) isn't transferred into or out of the system.
- **Isobaric** — Pressure (P) remains constant.
- **Isochoric** — Volume (V) remains constant.
- **Isothermal** — Temperature (T) remains constant.

In an **adiabatic** process, heat flow (Q) is zero. Applying the first law of thermodynamics, if ΔU=Q+W, and Q is 0, the change in internal energy of the gas must be equal to the work done on the gas (ΔU=W).

In an **isobaric** process, pressure of the gas remains constant. Because pressure is constant, the PV diagram for an isobaric process shows a horizontal line. Further, applying this to the ideal gas law, you find that V/T must remain constant for the process.

In an **isochoric** process, the volume of the gas remains constant. The PV diagram for an isochoric process is a vertical line. Because W=-PΔV, and ΔV=0, the work done on the gas is zero. This is also reflected graphically in the PV diagram. Work can be found by taking the area under the PV graph, but the area under a vertical line is zero. Applying this to the ideal gas law, you find that P/T must remain constant for an isochoric process.

In an **isothermal** process, temperature of the gas remains constant. Lines on a PV diagram describing any process held at constant temperature are therefore called isotherms. In an isothermal process, the product of the pressure and the volume of the gas remains constant. Further, because temperature is constant, the internal energy of the gas must remain constant.

9.31 Q: An ideal gas undergoes an adiabatic expansion, doing 2000 joules of work. How much does the gas's internal energy change?

9.31 A: Since the process is adiabatic, Q=0, therefore:

$$\Delta U = Q + W \xrightarrow{Q=0} \Delta U = W = -2000J$$

9.32 Q: Heat is removed from an ideal gas as its pressure drops from 200 kPa to 100 kPa. The gas then expands from a volume of 0.05 m³ to 0.1 m³ as shown in the PV diagram below. If curve AC represents an isotherm, find the work done by the gas and the heat added to the gas.

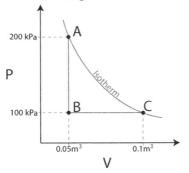

9.32 A: The work done by the gas in moving from A to B is zero, as the area under the graph is zero. In moving from B to C, however, the work done by the gas can be found by taking the area under the graph.

$$W = -P\Delta V = -(100,000\,Pa)(0.1m^3 - 0.05m^3) = -5000J$$

The negative sign indicates that 5000 joules of work was done by the gas. Since AC is on an isotherm, the temperature of the gas must remain constant, therefore the gas's internal energy must remain constant. Knowing that ΔU=Q+W, if ΔU=0, then Q must be equal to -W, therefore 5000 joules must have been added to the gas.

9.33 Q: One mole of an ideal gas undergoes a series of changes in pressure and volume as it follows path ABCA on the PV diagram. Using the PV diagram, find the temperature of the gas at each point, then fill in the information table by finding the change in internal energy, work done on, and heat added to the gas for each process.

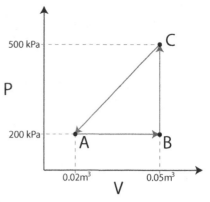

Process	ΔU	Q	W
A to B			
B to C			
C to A			

9.33 A: Start by finding the temperature at each point using PV=nRT.

$$PV = nRT \rightarrow T = \frac{PV}{nR}$$

$$T_A = \frac{P_A V_A}{nR} = \frac{(200,000\,Pa)(0.02m^3)}{1 \times 8.31\,{}^{J}\!/_{mol \bullet K}} = 481K$$

$$T_B = \frac{P_B V_B}{nR} = \frac{(200,000\,Pa)(0.05m^3)}{1 \times 8.31\,{}^{J}\!/_{mol \bullet K}} = 1203K$$

$$T_C = \frac{P_C V_C}{nR} = \frac{(500,000\,Pa)(0.05m^3)}{1 \times 8.31\,{}^{J}\!/_{mol \bullet K}} = 3008K$$

Next, fill in the table for isobaric process A to B.

$$W = -P\Delta V = -(200,000\,Pa)(0.05m^3 - 0.02m^3) = -6000J$$

$$\Delta U = \tfrac{3}{2}nR\Delta T = \tfrac{3}{2}nR(T_B - T_A) \rightarrow$$

$$\Delta U = \tfrac{3}{2}(1)(8.31\,{}^{J}\!/_{mol \bullet K})(1203K - 481K) = 9000J$$

$$\Delta U = Q + W \rightarrow Q = \Delta U - W \rightarrow$$

$$Q = 9000J - (-6000J) = 15000J$$

Now you can determine the work done in moving from B to C is zero because it is an isochoric (constant volume) process. Then, fill in the rest of the table for process B to C.

$$\Delta U = \tfrac{3}{2}nR\Delta T = \tfrac{3}{2}nR(T_C - T_B) \rightarrow$$

$$\Delta U = \tfrac{3}{2}(1)(8.31 \tfrac{J}{mol \bullet K})(3008K - 1203K) = 22{,}500J$$

$$\Delta U = Q + W \rightarrow Q = \Delta U - W \rightarrow$$
$$Q = 22{,}500J - 0 = 22{,}500J$$

Finally, you can determine the work done in moving from C to A by taking the area under the graph (noting that work is done on the gas as it is compressed, therefore it must be positive, then completing the remaining calculations in a similar fashion.

$$W = Area = \tfrac{1}{2}bh + lw = \tfrac{1}{2}(0.03m^3)(300000Pa) + 6000J$$

$$W = 10{,}500J$$

$$\Delta U = \tfrac{3}{2}nR\Delta T = \tfrac{3}{2}nR(T_A - T_C) \rightarrow$$

$$\Delta U = \tfrac{3}{2}(1)(8.31 \tfrac{J}{mol \bullet K})(481K - 3008K) = -31{,}500J$$

$$\Delta U = Q + W \rightarrow Q = \Delta U - W \rightarrow$$
$$Q = -31{,}500J - 10{,}500J = -42{,}000J$$

Process Summary Table			
Process	ΔU	Q	W
A to B	9,000	15,000	-6,000
B to C	22,500	22,500	0
C to A	-31,500	-42,000	10,500

The second law of thermodynamics can be stated in a variety of ways. One statement of this law says that heat flows naturally from a warmer object to a colder object, and cannot flow from a colder object to a warmer object without doing work on the system. This can be observed quite easily in everyday circumstances. For example, your cold spoon contacting your hot soup never results in your soup becoming hotter and your spoon becoming colder.

The second law of thermodynamics also limits the efficiency of any heat engine, and proves that it is not possible to make a 100 percent efficient heat engine, even if friction were completely eliminated.

Another statement of this law says that the level of entropy, or disorder, in a closed system can only increase or remain the same. This means that your desk will never naturally become more organized without doing work. It also means that you can't drop a handful of plastic building blocks and observe them spontaneously land in an impressive model of a medieval castle. Unfortunately, it even means that no matter how many times Humpty Dumpty falls off his wall, all his pieces on the ground will never end up more organized after he hits the ground compared to before his balance failed him.

The third and final law of thermodynamics, also known as Nernst's Theorem after its discoverer, Walter Nernst, states that no material can ever be cooled to absolute zero (although materials can get awfully close!)

Chapter 10: Electrostatics

"Electricity can be dangerous. My nephew tried to stick a penny into a plug. Whoever said a penny doesn't go far didn't see him shoot across that floor.

I told him he was grounded."

— Tim Allen

Objectives

1. Calculate the charge on an object.
2. Describe the differences between conductors and insulators.
3. Explain the difference between conduction and induction.
4. Explain how an electroscope works.
5. Solve problems using the law of conservation of charge.
6. Use Coulomb's Law to solve problems related to electrical force.
7. Recognize that objects that are charged exert forces, both attractive and repulsive.
8. Compare and contrast Newton's Law of Universal Gravitation with Coulomb's Law.
9. Define, measure, and calculate the strength of an electric field.
10. Solve problems related to charge, electric field, and forces.
11. Define and calculate electric potential energy.
12. Define and calculate potential difference.
13. Solve basic problems involving charged parallel plates.

Electricity and magnetism play a profound role in almost all aspects of life. From the moment you wake up, to the moment you go to sleep (and even while you're sleeping), applications of electricity and magnetism provide tools, light, warmth, transportation, communication, and even entertainment. Despite its widespread use, however, there is much about these phenomena that is not well understood.

Electric Charges

Matter is made up of atoms. Once thought to be the smallest building blocks of matter, scientists now know that atoms can be broken up into even smaller pieces, known as protons, electrons, and neutrons. Each atom consists of a dense core of positively charged protons and uncharged (neutral) neutrons. This core is known as the nucleus. It is surrounded by a "cloud" of much smaller, negatively charged electrons. These electrons orbit the nucleus in distinct energy levels. To move to a higher energy level, an electron must absorb energy. When an electron falls to a lower energy level, it gives off energy.

Most atoms are neutral -- that is, they have an equal number of positive and negative charges, giving a net charge of 0. In certain situations, however, an atom may gain or lose electrons. In these situations, the atom as a whole is no longer neutral and is called an **ion**. If an atom loses one or more electrons, it has a net positive charge and is known as a positive ion. If, instead, an atom gains one or more electrons, it has a net negative charge and is therefore called a negative ion. Like charges repel each other, while opposite charges attract each other. In physics, the charge on an object is represented with the symbol q.

Charge is a fundamental measurement in physics, much as length, time, and mass are fundamental measurements. The fundamental unit of charge is the Coulomb [C], which is a very large amount of charge. Compare that to the magnitude of charge on a single proton or electron, known as an elementary charge (e), which is equal to 1.6×10^{-19} coulomb. It would take 6.25×10^{18} elementary charges to make up a single coulomb of charge!

10.01 Q: An object possessing an excess of 6.0×10^6 electrons has what net charge?

10.01 A: $6 \times 10^6 \text{ electrons} \bullet \dfrac{-1.6 \times 10^{-19} C}{1 \text{ electron}} = -9.6 \times 10^{-13} C$

10.02 Q: An alpha particle consists of two protons and two neutrons. What is the charge of an alpha particle?

(1) 1.25×10^{19} C

(2) 2.00 C

(3) 6.40×10^{-19} C

(4) 3.20×10^{-19} C

10.02 A: (4) The net charge on the alpha particle is +2 elementary charges.

$$2e \bullet \frac{1.6 \times 10^{-19} C}{1e} = 3.2 \times 10^{-19} C$$

10.03 Q: If an object has a net negative charge of 4 coulombs, the object possesses

(1) 6.3×10^{18} more electrons than protons

(2) 2.5×10^{19} more electrons than protons

(3) 6.3×10^{18} more protons than electrons

(4) 2.5×10^{19} more protons than electrons

10.03 A: (2) $-4C \bullet \dfrac{1e}{1.6 \times 10^{-19} C} = -2.5 \times 10^{19} e$

10.04 Q: Which quantity of excess electric charge could be found on an object?

(1) 6.25×10^{-19} C

(2) 4.80×10^{-19} C

(3) 6.25 elementary charges

(4) 1.60 elementary charges

10.04 A: (2) all other choices require fractions of an elementary charge, while choice (2) is an integer multiple (3e) of elementary charges.

10.05 Q: What is the net electrical charge on a magnesium ion that is formed when a neutral magnesium atom loses two electrons?

(1) -3.2×10^{-19} C

(2) -1.6×10^{-19} C

(3) $+1.6 \times 10^{-19}$ C

(4) $+3.2 \times 10^{-19}$ C

10.05 A: (4) the net charge must be +2e, or 2(1.6×10^{-19} C)=3.2×10^{-19} C.

Conductors and Insulators

Certain materials allow electric charges to move freely. These are called **conductors**. Examples of good conductors include metals such as gold, copper, silver, and aluminum. In contrast, materials in which electric charges cannot move freely are known as **insulators**. Good insulators include materials such as glass, plastic, and rubber. Metals are better conductors of electricity compared to insulators because metals contain more free electrons.

Conductors and insulators are characterized by their resistivity, or ability to resist movement of charge. Materials with high resistivities are good insulators. Materials with low resistivities are good conductors.

Semiconductors are materials which, in pure form, are good insulators. However, by adding small amounts of impurities known as dopants, their resistivities can be lowered significantly until they become good conductors.

Charging by Conduction

Materials can be charged by contact, or **conduction**. If you take a balloon and rub it against your hair, some of the electrons from the atoms in your hair are transferred to the balloon. The balloon now has extra electrons, and therefore has a net negative charge. Your hair has a deficiency of electrons, therefore it now has a net positive charge.

Much like momentum and energy, charge is also conserved. Continuing the hair and balloon example, the magnitude of the net positive charge on your hair is equal to the magnitude of the net negative charge on the balloon. The total charge of the hair/balloon system remains zero (neutral). For every extra electron (negative charge) on the balloon, there is a corresponding missing electron (positive charge) in your hair. This known as the law of conservation of charge.

Conductors can also be charged by contact. If a charged conductor is brought into conduct with an identical neutral conductor, the net charge will be shared across the two conductors.

10.06 Q: If a conductor carrying a net charge of 8 elementary charges is brought into contact with an identical conductor with no net charge, what will be the charge on each conductor after they are separated?

10.06 A: Each conductor will have a charge of 4 elementary charges.

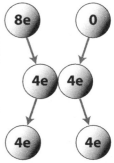

10.07 Q: What is the net charge (in coulombs) on each conductor after they are separated?

10.07 A: $q=4e=4(1.6\times10^{-19}$ C$)=6.4\times10^{-19}$ C

10.08 Q: Metal sphere A has a charge of −2 units and an identical metal sphere, B, has a charge of −4 units. If the spheres are brought into contact with each other and then separated, the charge on sphere B will be

(1) 0 units

(2) -2 units

(3) -3 units

(4) +4 units

10.08 A: (3) -3 units.

10.09 Q: Compared to insulators, metals are better conductors of electricity because metals contain more free

(1) protons

(2) electrons

(3) positive ions

(4) negative ions

10.09 A: (2) electrons.

A simple tool used to detect small electric charges known as an **electroscope** functions on the basis of conduction. The electroscope consists of a conducting rod attached to two thin conducting leaves at one end and isolated from surrounding charges by an insulating stopper placed in a flask. If a charged object is placed in contact with the conducting rod, part of the charge is transferred to the rod. Because the rod and leaves form a conducting path and like charges repel each other, the charges are distributed equally along the entire rod and leaf apparatus. The leaves, having like charges, repel each other, with larger charges providing greater leaf separation!

Charging by Induction

Conductors can also be charged without coming into contact with another charged object in a process known as charging by **induction**. This is accomplished by placing the conductor near a charged object and grounding the conductor. To understand charging by induction, you must first realize that when an object is connected to the earth by a conducting path, known as grounding, the earth acts like an infinite source for providing or accepting excess electrons.

To charge a conductor by induction, you first bring it close to another charged object. When the conductor is close to the charged object, any free electrons on the conductor will move toward the charged object if the object is positively charged (since opposite charges attract) or away from the charged object if the object is negatively charged (since like charges repel).

If the conductor is then "grounded" by means of a conducting path to the earth, the excess charge is compensated for by means of electron transfer to or from earth. Then the ground connection is severed. When the originally charged object is moved far away from the conductor, the charges in the conductor redistribute, leaving a net charge on the conductor as shown.

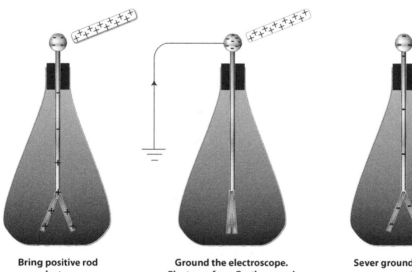

Bring positive rod
near electroscope.

Ground the electroscope.
Electrons from Earth ground
balance charge of positive rod.

Sever ground path and
remove positive rod.

You can also induce a charge in a charged region in a neutral object by bringing a strong positive or negative charge close to that object. In such cases, the electrons in the neutral object tend to move toward a strong positive charge, or away from a large negative charge. Though the object itself remains neutral, portions of the object are more positive or negative than other parts. In this way, you can attract a neutral object by bringing a charged object close to it, positive or negative. Put another way, a positively charged object can be attracted to both a negatively charged object and a neutral object, and a negatively charged object can be attracted to both a positively charged object and a neutral object.

For this reason, the only way to tell if an object is charged is by repulsion. A positively charge object can only be repelled by another positive charge and a negatively charged object can only be repelled by another negative charge.

10.10 Q: A positively charged glass rod attracts object X. The net charge of object X

(1) may be zero or negative

(2) may be zero or positive

(3) must be negative

(4) must be positive

10.10 A: (1) a positively charged rod can attract a neutral object or a negatively charged object.

10.11 Q: The diagram below shows three neutral metal spheres, x, y, and z, in contact and on insulating stands.

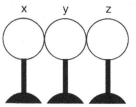

Which diagram best represents the charge distribution on the spheres when a positively charged rod is brought near sphere x, but does not touch it?

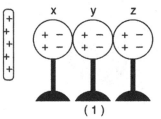

(1)

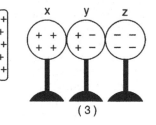

(3)

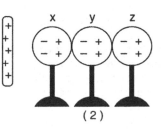

(2)

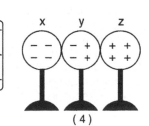

(4)

10.11 A: (4) is the correct answer.

Coulomb's Law

Like charges repel and opposite charges attract. In order for charges to repel or attract, they apply a force upon each either, known as the **electrostatic force**. Similar to the manner in which the force of attraction between two masses is determined by the amount of mass and the distance between the masses, as described by Newton's Law of Universal Gravitation, the force of attraction or repulsion is determined by the amount of charge and the distance between the charges.

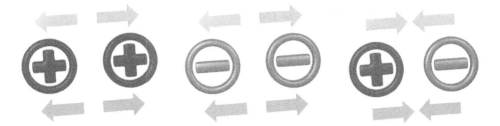

The magnitude of the electrostatic force is described by **Coulomb's Law**, which states that the magnitude of the electrostatic force (F_e) between two objects is equal to a constant, k, multiplied by each of the two charges, q_1 and q_2, and divided by the square of the distance between the charges (r^2). The constant k is known as the **electrostatic constant** and is given as k=8.99×10^9 N·m²/C².

$$F_e = \frac{kq_1 q_2}{r^2}$$

Notice how similar this formula is to the formula for the gravitational force! Both Newton's Law of Universal Gravitation and Coulomb's Law follow the inverse-square relationship, a pattern that repeats many times over in physics. The further you get from the charges, the weaker the electrostatic force. If you were to double the distance from a charge, you would quarter the electrostatic force on a charge.

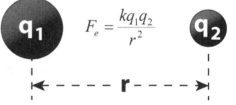

Formally, a positive value for the electrostatic force indicates that the force is a repelling force, while a negative value for the electrostatic force indicates that the force is an attractive force. Because force is a vector, you must assign a direction to it. To determine the direction of the force vector, once you have calculated its magnitude, use common sense to tell you the direction on each charged object. If the objects have opposite charges, they attract each other, and if they have like charges, they repel each other.

Chapter 10: Electrostatics

10.12 Q: Three protons are separated from a single electron by a distance of 1×10^{-6} m. Find the electrostatic force between them. Is this force attractive or repulsive?

10.12 A: $q_1 = 3 \text{ protons} = 3(1.6 \times 10^{-19} C) = 4.8 \times 10^{-19} C$

$q_2 = 1 \text{ electron} = 1(-1.6 \times 10^{-19} C) = -1.6 \times 10^{-19} C$

$$F_e = \frac{kq_1q_2}{r^2} = \frac{(8.99 \times 10^9 \ \frac{N \bullet m^2}{C^2})(4.8 \times 10^{-19} C)(-1.6 \times 10^{-19} C)}{(1 \times 10^{-6} m)^2}$$

$F_e = -6.9 \times 10^{-16} N$

10.13 Q: A distance of 1.0 meter separates the centers of two small charged spheres. The spheres exert gravitational force F_g and electrostatic force F_e on each other. If the distance between the spheres' centers is increased to 3.0 meters, the gravitational force and electrostatic force, respectively, may be represented as

(1) $F_g/9$ and $F_e/9$

(2) $F_g/3$ and $F_e/3$

(3) $3F_g$ and $3F_e$

(4) $9F_g$ and $9F_e$

10.13 A: (1) due to the inverse square law relationships.

10.14 Q: A beam of electrons is directed into the electric field between two oppositely charged parallel plates, as shown in the diagram below.

Electron beam

The electrostatic force exerted on the electrons by the electric field is directed

(1) into the page

(2) out of the page

(3) toward the bottom of the page

(4) toward the top of the page

10.14 A: (4) toward the top of the page because the electron beam is negative, and will be attracted by the positively charged upper plate and repelled by the negatively charged lower plate.

10.15 Q: The centers of two small charged particles are separated by a distance of 1.2×10^{-4} meter. The charges on the particles are $+8.0 \times 10^{-19}$ coulomb and $+4.8 \times 10^{-19}$ coulomb, respectively.

(A) Calculate the magnitude of the electrostatic force between these two particles.

(B) Sketch a graph showing the relationship between the magnitude of the electrostatic force between the two charged particles and the distance between the centers of the particles.

10.15 A: (A) $F_e = \dfrac{k q_1 q_2}{r^2} = \dfrac{(8.99 \times 10^9 \; \frac{N \bullet m^2}{C^2})(8.0 \times 10^{-19} C)(4.8 \times 10^{-19} C)}{(1.2 \times 10^{-4} m)^2}$

$F_e = 2.4 \times 10^{-19} N$

(B)

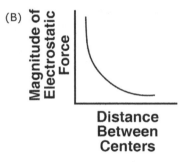

10.16 Q: The diagram below shows a beam of electrons fired through the region between two oppositely charged parallel plates in a cathode ray tube.

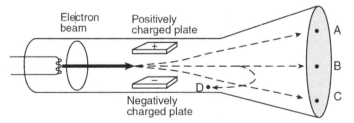

After passing between the charged plates, the electrons will most likely travel path

(1) A

(2) B

(3) C

(4) D

10.16 A: (1) A

Electric Fields

Similar to gravity, the electrostatic force is a non-contact force, or field force. Charged objects do not have to be in contact with each other to exert a force on each other. Somehow, a charged object feels the effect of another charged object through space. The property of space that allows a charged object to feel a force is a concept called the electric field. Although you cannot see an electric field, you can detect its presence by placing a positive test charge at various points in space and measuring the force the test charge feels.

While looking at gravity, the gravitational field strength was the amount of force observed by a mass per unit mass. In similar fashion, the electric field strength is the amount of electrostatic force observed by a charge per unit charge. Therefore, the electric field strength, E, is the electrostatic force observed at a given point in space divided by the test charge itself. Electric field strength is measured in Newtons per Coulomb (N/C).

$$E = \frac{F_e}{q}$$

10.17 Q: Two oppositely charged parallel metal plates, 1.00 centimeter apart, exert a force with a magnitude of 3.60×10^{-15} newtons on an electron placed between the plates. Calculate the magnitude of the electric field strength between the plates.

10.17 A: $E = \frac{F_e}{q} = \frac{3.6 \times 10^{-15}\,N}{1.6 \times 10^{-19}\,C} = 2.25 \times 10^4 \,{}^{N}\!/_{C}$

10.18 Q: Which quantity and unit are correctly paired?
(1) resistivity and Ω/m
(2) potential difference and eV
(3) current and C•s
(4) electric field strength and N/C

10.18 A: (4) electric field strength and N/C.

10.19 Q: What is the magnitude of the electric field intensity at a point where a proton experiences an electrostatic force of magnitude 2.30×10^{-25} newtons?
(1) 3.68×10^{-44} N/C
(2) 1.44×10^{-6} N/C
(3) 3.68×10^6 N/C
(4) 1.44×10^{44} N/C

10.19 A: (2) $E = \dfrac{F_e}{q} = \dfrac{2.3 \times 10^{-25}\,N}{1.6 \times 10^{-19}\,C} = 1.44 \times 10^{-6}\,{}^{N}\!/_{C}$

10.20 Q: The diagram below represents an electron within an electric field between two parallel plates that are charged with a potential difference of 40.0 volts.

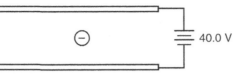

If the magnitude of the electric force on the electron is 2.00×10^{-15} newtons, the magnitude of the electric field strength between the charged plates is

(1) 3.20×10^{-34} N/C

(2) 2.00×10^{-14} N/C

(3) 1.25×10^{4} N/C

(4) 2.00×10^{16} N/C

10.20 A: (3) $E = \dfrac{F_e}{q} = \dfrac{2 \times 10^{-15}\,N}{1.6 \times 10^{-19}\,C} = 1.25 \times 10^{4}\,{}^{N}\!/_{C}$

Since you can't actually see the electric field, you can draw electric field lines to help visualize the force a charge would feel if placed at a specific position in space. These lines show the direction of the electric force a positively charged particle would feel at that point. The more dense the lines are, the stronger the force a charged particle would feel, therefore the stronger the electric field. As the lines get further apart, the strength of the electric force a charged particle would feel is smaller, therefore the electric field is smaller.

By convention, electric field lines are drawn showing the direction of force on a positive charge. Therefore, to draw electric field lines for a system of charges, follow these basic rules:

1. Electric field lines point away from positive charges and toward negative charges.
2. Electric field lines never cross.
3. Electric field lines always intersect conductors at right angles to the surface.
4. Stronger fields have closer lines.
5. Field strength and line density decreases as you move away from the charges.

Let's take a look at a few examples of electric field lines, starting with isolated positive (left) and negative (right) charges. Notice that for each charge, the lines radiate outward or inward spherically. The lines point away from the positive charge, since a positive test charge placed in the field (near the fixed charge) would feel a repelling force. The lines point in toward the negative fixed charge, since a positive test charge would feel an attractive force.

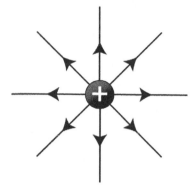 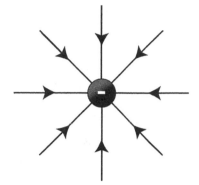

If you have both positive and negative charges in close proximity, you follow the same basic procedure:

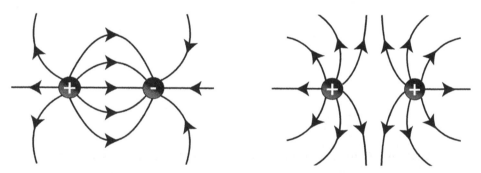

10.21 Q: Two small metallic spheres, A and B, are separated by a distance of 4.0×10^{-1} meter, as shown. The charge on each sphere is $+1.0 \times 10^{-6}$ coulomb. Point P is located near the spheres.

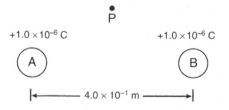

(A) What is the magnitude of the electrostatic force between the two charged spheres?

(1) 2.2×10^{-2} N

(2) 5.6×10^{-2} N

(3) 2.2×10^{4} N

(4) 5.6×10^{4} N

(B) Which arrow best represents the direction of the resultant electric field at point P due to the charges on spheres A and B?

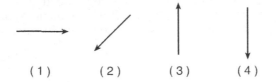

(1)   (2)   (3)   (4)

10.21 A: (A) 2 $F_e = \dfrac{kq_1 q_2}{r^2} = \dfrac{(8.99 \times 10^9 \frac{N \cdot m^2}{C^2})(1.0 \times 10^{-6} C)(1.0 \times 10^{-6} C)}{(4 \times 10^{-1} m)^2}$

$F_e = 0.056N$

(B) Correct answer is 3.

10.22 Q: In the diagram below, P is a point near a negatively charged sphere.

Which vector best represents the direction of the electric field at point P?

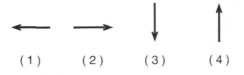

(1) (2) (3) (4)

10.22 A: Correct answer is (1). Electric field lines point in toward negative charges.

10.23 Q: Sketch at least four electric field lines with arrowheads that represent the electric field around a negatively charged conducting sphere.

10.23 A:

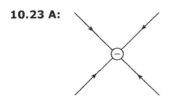

10.24 Q: The centers of two small charged particles are separated by a distance of 1.2×10^{-4} meter. The charges on the particles are $+8.0\times10^{-19}$ coulomb and $+4.8\times10^{-19}$ coulomb, respectively. Sketch at least four electric field lines in the region between the two positively charged particles.

10.24 A:

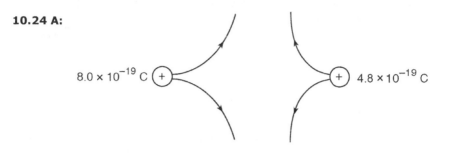

10.25 Q: Which graph best represents the relationship between the magnitude of the electric field strength, E, around a point charge and the distance, r, from the point charge?

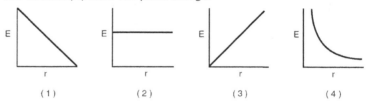

10.25 A: Correct answer is 4.

Because gravity and electrostatics have so many similarities, let's take a minute to do a quick comparison of electrostatics and gravity.

Electrostatics	Gravity
Force: $F_e = \dfrac{kq_1q_2}{r^2}$	**Force:** $F_g = \dfrac{Gm_1m_2}{r^2}$
Field Strength: $E = \dfrac{F_e}{q}$	**Field Strength:** $g = \dfrac{F_g}{m}$
Field Strength: $E = \dfrac{kq}{r^2}$	**Field Strength:** $g = \dfrac{Gm}{r^2}$
Constant: k=8.99×10^9 N·m²/C²	**Constant:** G=6.67×10^{-11} N·m²/kg²
Charge Units: Coulombs	**Mass Units:** kilograms

What is the big difference between electrostatics and gravity? The gravitational force can only attract, while the electrostatic force can both attract and repel. Notice again that both the electric field strength and the gravitational field strength follow the inverse-square law relationship. Field strength is inversely related to the square of the distance.

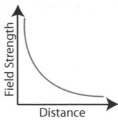

Electric Potential Difference

When an object was lifted against the force of gravity by applying a force over a distance, work was done to give that object gravitational potential energy. The same concept applies to electric fields. If you move a charge against an electric field, you must apply a force for some distance, therefore you do work and give it electrical potential energy. The work done per unit charge in moving a charge between two points in an electric field is known as the **electric potential difference**, (V). The units of electric potential are volts, where a volt is equal to 1 Joule per Coulomb. Therefore, if you do 1 Joule of work in moving a charge of 1 Coulomb in an electric field, the electric potential difference between those points would be 1 volt. This is described mathematically by:

$$V = \frac{W}{q}$$

V in this formula is potential difference (in volts), W is work or electrical energy (in Joules), and q is your charge (in Coulombs). Let's take a look at some sample problems.

10.26 Q: A potential difference of 10 volts exists between two points, A and B, within an electric field. What is the magnitude of charge that requires 2.0×10^{-2} joules of work to move it from A to B?

10.26 A: $V = \frac{W}{q}$

$$q = \frac{W}{V} = \frac{2 \times 10^{-2} J}{10 V} = 2 \times 10^{-3} C$$

10.27 Q: How much electrical energy is required to move a 4.00-microcoulomb charge through a potential difference of 36.0 volts?

(1) 9.00×10^6 J

(2) 144 J

(3) 1.44×10^{-4} J

(4) 1.11×10^{-7} J

10.27 A: (3) $V = \dfrac{W}{q}$

$$W = qV = (4 \times 10^{-6} C)(36V) = 1.44 \times 10^{-4} J$$

10.28 Q: If 1.0 joule of work is required to move 1.0 coulomb of charge between two points in an electric field, the potential difference between the two points is

(1) 1.0×10^0 V

(2) 9.0×10^9 V

(3) 6.3×10^{18} V

(4) 1.6×10^{-19} V

10.28 A: (1) $V = \dfrac{W}{q} = \dfrac{1J}{1C} = 1V = 1.0 \times 10^0 V$

10.29 Q: If 60 joules of work is required to move 5 coulombs of charge between two points in an electric field, what is the potential difference between these points?

(1) 5 V

(2) 12 V

(3) 60 V

(4) 300 V

10.29 A: (2) $V = \dfrac{W}{q} = \dfrac{60J}{5C} = 12V$

10.30 Q: In an electric field, 0.90 joules of work is required to bring 0.45 coulombs of charge from point A to point B. What is the electric potential difference between points A and B?

(1) 5.0 V

(2) 2.0 V

(3) 0.50 V

(4) 0.41 V

10.30 A: (2) $V = \dfrac{W}{q} = \dfrac{0.90J}{0.45C} = 2V$

When dealing with electrostatics, often times the amount of electric energy or work done on a charge is a very small portion of a Joule. Dealing with such small numbers is cumbersome, so physicists devised an alternate unit for electrical energy and work that can be more convenient than the Joule. This unit, known as the electronvolt (eV), is the amount of work done in moving an elementary charge through a potential difference of 1V. One electron-volt, therefore, is equivalent to one volt multiplied by one elementary charge (in Coulombs): 1 eV = 1.6×10^{-19} Joules.

10.31 Q: A proton is moved through a potential difference of 10 volts in an electric field. How much work, in electronvolts, was required to move this charge?

10.31 A: $V = \dfrac{W}{q}$

$W = qV = (1e)(10V) = 10eV$

Parallel Plates

If you know the potential difference between two parallel plates, you can easily calculate the electric field strength between the plates. As long as you're not near the edge of the plates, the electric field is constant between the plates and its strength is given by the equation:

$$E = \frac{V}{d}$$

You'll note that with the potential difference V in volts, and the distance between the plates in meters, units for the electric field strength are volts per meter [V/m]. Previously, the units for electric field strength were given as newtons per Coulomb [N/C]. It is easy to show these are equivalent:

$$\frac{N}{C} = \frac{N \bullet m}{C \bullet m} = \frac{J}{C \bullet m} = \frac{J/C}{m} = \frac{V}{m}$$

10.32 Q: The magnitude of the electric field strength between two oppositely charged parallel metal plates is 2.0×10^3 newtons per coulomb. Point P is located midway between the plates.

(A) Sketch at least five electric field lines to represent the field between the two oppositely charged plates.

(B) An electron is located at point P between the plates. Calculate the magnitude of the force exerted on the electron by the electric field.

10.32 A: (A)

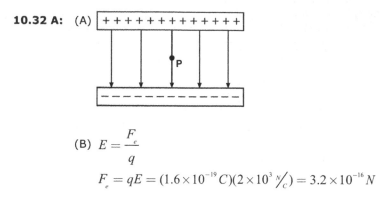

(B) $E = \dfrac{F_e}{q}$

$$F_e = qE = (1.6 \times 10^{-19}\,C)(2 \times 10^3\,\sqrt[N]{c}) = 3.2 \times 10^{-16}\,N$$

10.33 Q: A moving electron is deflected by two oppositely charged parallel plates, as shown in the diagram below.

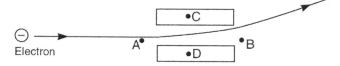

The electric field between the plates is directed from

(1) A to B

(2) B to A

(3) C to D

(4) D to C

10.33 A: (3) C to D because the electron feels a force opposite the direction of the electric field due to its negative charge.

10.34 Q: An electron is located in the electric field between two parallel metal plates as shown in the diagram below.

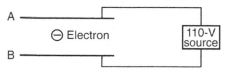

If the electron is attracted to plate A, then plate A is charged

(1) positively, and the electric field runs from plate A to plate B

(2) positively, and the electric field runs from plate B to plate A

(3) negatively, and the electric field runs from plate A to plate B

(4) negatively, and the electric field runs from plate B to plate A

10.34 A: Correct answer is 1.

10.35 Q: An electron placed between oppositely charged parallel plates moves toward plate A, as represented in the diagram below.

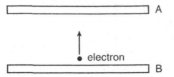

What is the direction of the electric field between the plates?

(1) toward plate A

(2) toward plate B

(3) into the page

(4) out of the page

10.35 A: (2) toward plate B.

10.36 Q: The diagram below represents two electrons, e_1 and e_2, located between two oppositely charged parallel plates.

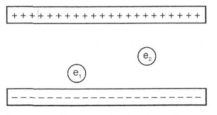

Compare the magnitude of the force exerted by the electric field on e_1 to the magnitude of the force exerted by the electric field on e_2.

10.36 A: The forces are the same because the electric field between two parallel plates is constant.

Capacitors

Parallel conducting plates separated by an insulator can be used to store electrical charge. These devices come in a variety of sizes, and are known as parallel plate capacitors. The amount of charge a capacitor can store on a single plate for a given amount of potential difference across the plates is known as the device's capacitance, given in coulombs per volt, also known as a Farad (F). A Farad is a very large amount of capacitance, therefore most capacitors have values in the microFarad, nanoFarad, and even pico-Farad ranges.

$$C = \frac{q}{V}$$

10.37 Q: A capacitor stores 3 microcoulombs of charge with a potential difference of 1.5 volts across the plates. What is the capacitance?

10.37 A: $C = \frac{q}{V} = \frac{3 \times 10^{-6} C}{1.5V} = 2 \times 10^{-6} F$

10.38 Q: How much charge sits on the top plate of a 200 nF capacitor when charged to a potential difference of 6 volts?

10.38 A: $C = \frac{q}{V} \rightarrow q = CV = (200 \times 10^{-9} F)(6V) = 1.2 \times 10^{-6} C$

The amount of charge a capacitor can hold is determined by its geometry as well as the insulating material between the plates. The capacitance is direct-ly related to the area of the plates, and inversely related to the separation between the plates, as shown in the formula below. The permittivity of an insulator (ε) describes the insulator's resistance to the creation of an electric field, and is equal to 8.85×10⁻¹² Farads per meter for an air-gap capacitor.

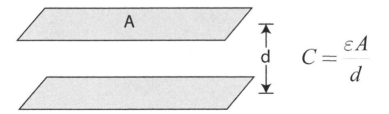

10.39 Q: Find the capacitance of two parallel plates of length 1 millimeter and width 2 millimeters if they are separated by 3 micrometers of air.

10.39 A: $C = \frac{\varepsilon A}{d} = \frac{(8.85 \times 10^{-12} \, F/m)(.001m \times .002m)}{3 \times 10^{-6} m} = 5.9 \times 10^{-12} F$

10.40 Q: Find the distance between the plates of a 5 nanoFarad air-gap capacitor with a plate area of 0.06 m².

10.40 A:
$$C = \frac{\varepsilon A}{d} \rightarrow d = \frac{\varepsilon A}{C} \rightarrow$$
$$d = \frac{(8.85 \times 10^{-12} \, ^{F}\!/_{m})(0.06m^2)}{5 \times 10^{-9} \, F} = 1.1 \times 10^{-4} \, m$$

Equipotential Lines

Much like looking at a topographic map which shows you lines of equal altitude, or equal gravitational potential energy, you can make a map of the electric field and connect points of equal electrical potential. These lines, known as **equipotential lines**, always cross electrical field lines at right angles, and show positions in space with constant electrical potential. If you move a charged particle in space, and it always stays on an equipotential line, no work will be done.

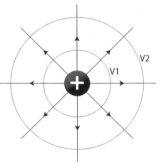

Chapter 11: Current Electricity

"And God said, 'Let there be light' and there was light, but the electricity board said He would have to wait until Thursday to be connected."

— Spike Milligan

Objectives

1. Define and calculate electric current.
2. Define and calculate resistance using Ohm's law.
3. Explain the factors and calculate the resistance of a conductor.
4. Identify the path and direction of current flow in a circuit.
5. Draw and interpret schematic diagrams of circuits.
6. Effectively use and analyze voltmeters and ammeters.
7. Solve series and parallel circuit problems using VIRP tables.
8. Calculate equivalent resistances for resistors in both series and parallel configurations.
9. Calculate power and energy used in electric circuits.

Electric Current

Electric current is the flow of charge, much like water currents are the flow of water molecules. Water molecules tend to flow from areas of high gravitational potential energy to low gravitational potential energy. Electric currents flow from high electric potential to low electric potential. The greater the difference between the high and low potential, the more current that flows!

In a majority of electric currents, the moving charges are negative electrons. However, due to historical reasons dating back to Ben Franklin, we say that conventional current flows in the direction positive charges would move. Although inconvenient, it's fairly easy to keep straight if you just remember that the actual moving charges, the electrons, flow in a direction opposite that of the electric current. With this in mind, you can state that positive current flows from high potential to low potential, even though the charge carriers (electrons) actually flow from low to high potential.

Electric current (I) is measured in amperes (A), or amps, and can be calculated by finding the total amount of charge (Δq), in coulombs, which passes a specific point in a given time (t). Electric current can therefore be calculated as:

$$I = \frac{\Delta q}{t}$$

11.01 Q: A charge of 30 Coulombs passes through a 24-ohm resistor in 6.0 seconds. What is the current through the resistor?

11.01 A: $I = \frac{\Delta q}{t} = \frac{30C}{6s} = 5A$

11.02 Q: Charge flowing at the rate of 2.50×10^{16} elementary charges per second is equivalent to a current of
(1) 2.50×10^{13} A
(2) 6.25×10^{5} A
(3) 4.00×10^{-3} A
(4) 2.50×10^{-3} A

11.02 A: $I = \frac{\Delta q}{t} = \frac{(2.50 \times 10^{16})(1.6 \times 10^{-19}C)}{1s} = 4 \times 10^{-3}A$

11.03 Q: The current through a lightbulb is 2.0 amperes. How many coulombs of electric charge pass through the lightbulb in one minute?
(1) 60 C
(2) 2.0 C

(3) 120 C

(4) 240 C

11.03 A: (3) $I = \dfrac{\Delta q}{t}$

$\Delta q = It = (2\,A)(60s) = 120C$

11.04 Q: A 1.5-volt, AAA cell supplies 750 milliamperes of current through a flashlight bulb for 5 minutes, while a 1.5-volt, C cell supplies 750 milliamperes of current through the same flashlight bulb for 20 minutes. Compared to the total charge transferred by the AAA cell through the bulb, the total charge transferred by the C cell through the bulb is

(1) half as great

(2) twice as great

(3) the same

(4) four times as great

11.04 A: (4) If Δq=It, and both cells supply 0.750A but the C cell supplies the same current for four times as long, it must supply four times the total charge compared to the AAA cell.

11.05 Q: The current traveling from the cathode to the screen in a television picture tube is 5.0×10^{-5} amperes. How many electrons strike the screen in 5.0 seconds?

(1) 3.1×10^{24}

(2) 6.3×10^{18}

(3) 1.6×10^{15}

(4) 1.0×10^{5}

11.05 A: (3)

$I = \dfrac{\Delta q}{t}$

$\Delta q = It = (5 \times 10^{-5}\,A)(5s) = 2.5 \times 10^{-4}C$

$2.5 \times 10^{-4}C \bullet \dfrac{1\ \text{electron}}{1.6 \times 10^{-19}C} = 1.6 \times 10^{15}\ \text{electrons}$

Resistance

Electrical charges can move easily in some materials (conductors) and less freely in others (insulators). A material's ability to conduct electric charge is known as its **conductivity**. Good conductors have high conductivities. The conductivity of a material depends on:

1. Density of free charges available to move
2. Mobility of those free charges

In similar fashion, material's ability to resist the movement of electric charge is known as its **resistivity**, symbolized with the Greek letter rho (ρ). Resistivity is measured in ohm-meters, which is represented by the Greek letter omega multiplied by meters (Ω•m). Both conductivity and resistivity are properties of a material.

When an object is created out of a material, the material's tendency to conduct electricity, or conductance, depends on the material's conductivity as well as the material's shape. For example, a hollow cylindrical pipe has a higher conductivity of water than a cylindrical pipe filled with cotton. However, the shape of the pipe also plays a role. A very thick but short pipe can conduct lots of water, yet a very narrow, very long pipe can't conduct as much water. Both geometry of the object and the object's composition influence its conductance.

Focusing on an object's ability to resist the flow of electrical charge, objects made of high resistivity materials tend to impede electrical current flow and have a high resistance. Further, materials shaped into long, thin objects also increase an object's electrical resistance. Finally, objects typically exhibit higher resistivities at higher temperatures. You must take all of these factors into account together to describe an object's resistance to the flow of electrical charge. Resistance is a functional property of an object that describes the object's ability to impede the flow of charge through it. Units of resistance are ohms (Ω).

For any given temperature, you can calculate an object's electrical resistance, in ohms, using the following formula.

$$R = \frac{\rho L}{A}$$

Resistivities at 20°C	
Material	Resistivity (Ω•m)
Aluminum	2.82×10^{-8}
Copper	1.72×10^{-8}
Gold	2.44×10^{-8}
Nichrome	$150. \times 10^{-8}$
Silver	1.59×10^{-8}
Tungsten	5.60×10^{-8}

In this formula, R is the resistance of the object, in ohms (Ω), rho (ρ) is the resistivity of the material the object is made out of, in ohm•meters (Ω•m), L is the length of the object, in meters, and A is the cross-sectional area of the object, in meters squared. Note that a table of material resistivities for a constant temperature is given to you on the previous page as well.

11.06 Q: A 3.50-meter length of wire with a cross-sectional area of 3.14×10^{-6} m^2 at 20° Celsius has a resistance of 0.0625 Ω. Determine the resistivity of the wire and the material it is made out of.

11.06 A: $R = \dfrac{\rho L}{A}$

$$\rho = \frac{RA}{L} = \frac{(.0625\Omega)(3.14 \times 10^{-6}\, m^2)}{3.5m} = 5.6 \times 10^{-8}\,\Omega \bullet m$$

Material must be tungsten.

11.07 Q: The electrical resistance of a metallic conductor is inversely proportional to its

(1) temperature
(2) length
(3) cross-sectional area
(4) resistivity

11.07 A: (3) straight from the formula.

11.08 Q: At 20°C, four conducting wires made of different materials have the same length and the same diameter. Which wire has the least resistance?

(1) aluminum
(2) gold
(3) nichrome
(4) tungsten

11.08 A: (2) gold because it has the lowest resistivity.

11.09 Q: A length of copper wire and a 1.00-meter-long silver wire have the same cross-sectional area and resistance at 20°C. Calculate the length of the copper wire.

11.09 A: $R = \left(\dfrac{\rho L}{A}\right)_{copper} = \left(\dfrac{\rho L}{A}\right)_{silver}$

$$R = \dfrac{\rho_{copper} L_{copper}}{A} = \dfrac{\rho_{silver} L_{silver}}{A}$$

$$L_{copper} = \dfrac{\rho_{silver} L_{silver}}{\rho_{copper}} = \dfrac{(1.59 \times 10^{-8}\,\Omega m)(1m)}{1.72 \times 10^{-8}\,\Omega m}$$

$$L_{copper} = 0.924m$$

11.10 Q: A 10-meter length of copper wire is at 20°C. The radius of the wire is 1.0×10⁻³ meter.

Cross Section of Copper Wire

r = 1.0 × 10⁻³ m

(A) Determine the cross-sectional area of the wire.

(B) Calculate the resistance of the wire.

11.10 A: (A) $Area_{circle} = \pi r^2 = \pi(1.0 \times 10^{-3}\,m)^2 = 3.14 \times 10^{-6}\,m^2$

(B) $R = \dfrac{\rho L}{A} = \dfrac{(1.72 \times 10^{-8}\,\Omega m)(10m)}{3.14 \times 10^{-6}\,m^2} = 5.5 \times 10^{-2}\,\Omega$

Ohm's Law

If resistance opposes current flow, and potential difference promotes current flow, it only makes sense that these quantities must somehow be related. George Ohm studied and quantified these relationships for conductors and resistors in a famous formula now known as **Ohm's Law**:

$$R = \dfrac{V}{I}$$

Ohm's Law may make more qualitative sense if it is re-arranged slightly:

$$I = \dfrac{V}{R}$$

Now it's easy to see that the current flowing through a conductor or resistor (in amps) is equal to the potential difference across the object (in volts) divided by the resistance of the object (in ohms). If you want a large current to flow, you require a large potential difference (such as a large battery), and/or a very small resistance.

11.11 Q: The current in a wire is 24 amperes when connected to a 1.5 volt battery. Find the resistance of the wire.

11.11 A: $R = \dfrac{V}{I} = \dfrac{1.5V}{24A} = 0.0625\Omega$

11.12 Q: In a simple electric circuit, a 24-ohm resistor is connected across a 6-volt battery. What is the current in the circuit?

(1) 1.0 A

(2) 0.25 A

(3) 140 A

(4) 4.0 A

11.12 A: (2) $I = \dfrac{V}{R} = \dfrac{6V}{24\Omega} = 0.25A$

11.13 Q: A constant potential difference is applied across a variable resistor held at constant temperature. Which graph best represents the relationship between the resistance of the variable resistor and the current through it?

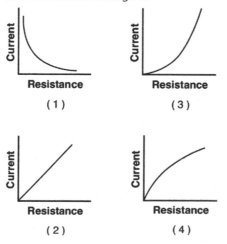

(1)   (3)

(2)   (4)

11.13 A: (1) due to Ohm's Law (I=V/R).

11.14 Q: What is the current in a 100-ohm resistor connected to a 0.40-volt source of potential difference?

(1) 250 mA

(2) 40 mA

(3) 2.5 mA

(4) 4.0 mA

11.14 A: (4) $I = \dfrac{V}{R} = \dfrac{0.40V}{100\Omega} = 0.004\,A = 4mA$

Note: Ohm's Law isn't truly a law of physics -- not all materials obey this relationship. It is, however, a very useful empirical relationship that accurately describes key electrical characteristics of conductors and resistors. One way to test if a material is ohmic (if it follows Ohm's Law) is to graph the voltage vs. current flow through the material. If the material obeys Ohm's Law, you get a linear relationship, where the slope of the line is equal to the material's resistance.

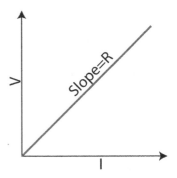

11.15 Q: The graph below represents the relationship between the potential difference (V) across a resistor and the current (I) through the resistor.

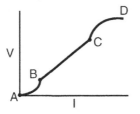

Through which entire interval does the resistor obey Ohm's law?

(1) AB

(2) BC

(3) CD

(4) AD

11.15 A: (2) BC because the graph is linear in this interval.

Electrical Circuits

An **electrical circuit** is a closed loop path through which current can flow. An electrical circuit can be made up of almost any materials (including humans if they're not careful), but practically speaking, circuits are typically comprised of electrical devices such as wires, batteries, resistors, and switches. Conventional current will flow through a complete closed loop (closed circuit) from high potential to low potential. Therefore, electrons actually flow in the opposite direction, from low potential to high potential. If the path isn't a closed loop (and is, instead, an open circuit), no current will flow.

Electric circuits, which are three-dimensional constructs, are typically represented in two dimensions using diagrams known as **circuit schematics**.

Circuit Symbols	
⊥	cell
⊥	battery
／_	switch
─(V)─	voltmeter
─(A)─	ammeter
⋀⋁⋀	resistor
⋀⋁⋀	variable resistor
─(ell)─	lamp

These schematics are simplified, standardized representations in which common circuit elements are represented with specific symbols, and wires connecting the elements in the circuit are represented by lines. Basic circuit schematic symbols are shown in the diagram at left.

In order for current to flow through a circuit, you must have a source of potential difference. Typical sources of potential difference are voltaic cells, batteries (which are just two or more cells connected together), and power (voltage) supplies. Voltaic cells are often referred to as batteries in common terminology. In drawing a cell or battery on a circuit schematic, remember that the longer side of the symbol is the positive terminal.

AA Voltaic Cell   9V Battery

Electric circuits must form a complete conducting path in order for current to flow. In the example circuit shown below left, the circuit is incomplete because the switch is open, therefore no current will flow and the lamp will not light. In the circuit below right, however, the switch is closed, creating a closed loop path. Current will flow and the lamp will light up.

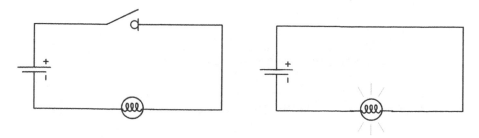

Note that in the picture at right, conventional current will flow from positive to negative, creating a clockwise current path in the circuit. The actual electrons in the wire, however, are flowing in the opposite direction, or counter-clockwise.

Energy & Power

Just like mechanical power is the rate at which mechanical energy is expended, **electrical power** is the rate at which electrical energy is expended. When you do work on something you change its energy. Further, electrical work or energy is equal to charge times potential difference. Therefore, you can combine these to write the equation for electrical power as:

$$P = \frac{W}{t} = \frac{qV}{t}$$

The amount of charge moving past a point per given unit of time is current, therefore you can continue the derivation as follows:

$$P = \left(\frac{q}{t}\right)V = IV$$

So electrical power expended in a circuit is the electrical current multiplied by potential difference (voltage). Using Ohm's Law, you can expand this even further to provide several different methods for calculating electrical power dissipated by a resistor:

$$P = VI = I^2R = \frac{V^2}{R}$$

Of course, conservation of energy still applies, so the energy used in the resistor is converted into heat (in most cases) and light, or it can be used to do work. Let's put this knowledge to use in a practical application.

11.16 Q: A 110-volt toaster oven draws a current of 6 amps on its highest setting as it converts electrical energy into thermal energy. What is the toaster's maximum power rating?

11.16 A: $P = VI = (110V)(6A) = 660W$

11.17 Q: An electric iron operating at 120 volts draws 10 amperes of current. How much heat energy is delivered by the iron in 30 seconds?

(1) 3.0×10^2 J

(2) 1.2×10^3 J

(3) 3.6×10^3 J

(4) 3.6×10^4 J

11.17 A: (4) $W = Pt = VIt = (120V)(10A)(30s) = 3.6 \times 10^4 J$

11.18 Q: One watt is equivalent to one

(1) N·m

(2) N/m

(3) J·s

(4) J/s

11.18 A: (4) J/s, since Power is W/t, and the unit of work is the joule, and the unit of time is the second.

11.19 Q: A potential drop of 50 volts is measured across a 250-ohm resistor. What is the power developed in the resistor?

(1) 0.20 W

(2) 5.0 W

(3) 10 W

(4) 50 W

11.19 A: $P = \dfrac{V^2}{R} = \dfrac{(50V)^2}{250\Omega} = 10W$

11.20 Q: What is the minimum information needed to determine the power dissipated in a resistor of unknown value?

(1) potential difference across the resistor, only

(2) current through the resistor, only

(3) current and potential difference, only

(4) current, potential difference, and time of operation

11.20 A: (3) current and potential difference, only (P=VI).

Voltmeters

Voltmeters are tools used to measure the potential difference between two points in a circuit. The voltmeter is connected in parallel with the element to be measured, meaning an alternate current path around the element to be measured and through the voltmeter is created. You have connected a voltmeter correctly if you can remove the voltmeter from the circuit without breaking the circuit. In the diagram at right, a voltmeter is connected to correctly measure the potential difference across the lamp. Voltmeters have very high resistance so as to minimize the current flow through the voltmeter and the voltmeter's impact on the circuit.

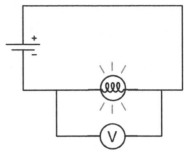

Ammeters

Ammeters are tools used to measure the current in a circuit. The ammeter is connected in series with the circuit, so that the current to be measured flows directly through the ammeter. The circuit must be broken to correctly insert an ammeter. Ammeters have very low resistance to minimize the potential drop through the ammeter and the ammeter's impact on the circuit, so inserting an ammeter into a circuit in parallel can result in extremely high currents and may destroy the ammeter. In the diagram at right, an ammeter is connected correctly to measure the current flowing through the circuit.

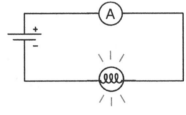

11.21 Q: In the electric circuit diagram, possible locations of an ammeter and voltmeter are indicated by circles 1, 2, 3, and 4. Where should an ammeter be located to correctly measure the total current and where should a voltmeter be located to correctly measure the total voltage?

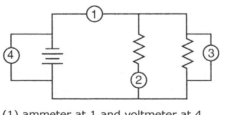

(1) ammeter at 1 and voltmeter at 4

(2) ammeter at 2 and voltmeter at 3

(3) ammeter at 3 and voltmeter at 4

(4) ammeter at 1 and voltmeter at 2

11.21 A: (1) To measure the total current, the ammeter must be placed at position 1, as all the current in the circuit must pass through this wire, and ammeters are always connected in series. To measure the total voltage in the circuit, the voltmeter could be placed at either position 3 or position 4. Voltmeters are always placed in parallel with the circuit element being analyzed, and positions 3 and 4 are equivalent because they are connected with wires (and potential is always the same anywhere in an ideal wire).

11.22 Q: Which circuit diagram below correctly shows the connection of ammeter A and voltmeter V to measure the current through and potential difference across resistor R?

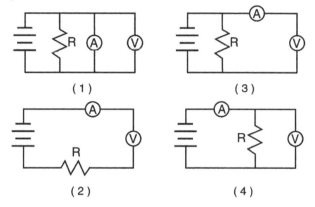

11.22 A: (4) shows an ammeter in series and a voltmeter in parallel with the resistor.

11.23 Q: A student uses a voltmeter to measure the potential difference across a resistor. To obtain a correct reading, the student must connect the voltmeter

(A) in parallel with the resistor

(B) in series with the resistor

(C) before connecting the other circuit components

(D) after connecting the other circuit components

11.23 A: (A) in parallel with the resistor.

11.24 Q: Which statement about ammeters and voltmeters is correct?

(1) The internal resistance of both meters should be low.

(2) Both meters should have a negligible effect on the circuit being measured.

(3) The potential drop across both meters should be made as large as possible.

(4) The scale range on both meters must be the same.

11.24 A: (2) Both meters should have a negligible effect on the circuit being measured.

11.25 Q: Compared to the resistance of the circuit being measured, the internal resistance of a voltmeter is designed to be very high so that the meter will draw

(1) no current from the circuit

(2) little current from the circuit

(3) most of the current from the circuit

(4) all the current from the circuit

11.25 A: (2) the voltmeter should draw as little current as possible from the circuit to minimize its effect on the circuit, but it does require some small amount of current to operate.

Series Circuits

Developing an understanding of circuits is the first step in learning about the modern-day electronic devices that dominate what is becoming known as the "Information Age." A basic circuit type, the **series circuit**, is a circuit in which there is only a single current path. Kirchhoff's Laws provide the tools in order to analyze any type of circuit.

Kirchhoff's Current Law (KCL), named after German physicist Gustav Kirchhoff, states that the sum of all current entering any point in a circuit has to equal the sum of all current leaving any point in a circuit. More simply, this is another way of looking at the law of conservation of charge.

Kirchhoff's Voltage Law (KVL) states that the sum of all the potential drops in any closed loop of a circuit has to equal zero. More simply, KVL is a method of applying the law of conservation of energy to a circuit.

11.26 Q: A 3.0-ohm resistor and a 6.0-ohm resistor are connected in series in an operating electric circuit. If the current through the 3.0-ohm resistor is 4.0 amperes, what is the potential difference across the 6.0-ohm resistor?

11.26 A: First, draw a picture of the situation. If 4 amps of current is flowing through the 3-ohm resistor, then 4 amps of current must be flowing through the 6-ohm resistor according to Kirchhoff's Current Law. Since you know the current and the resistance, you can calculate the voltage drop across the 6-ohm resistor using Ohm's Law: $V=IR=(4A)(6\Omega)=24V$.

11.27 Q: The diagram below represents currents in a segment of an electric circuit.

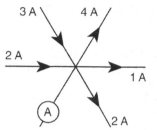

3 A   4 A

2 A

1 A

A

2 A

What is the reading of ammeter A?

(1) 1 A

(2) 2 A

(3) 3 A

(4) 4 A

11.27 A: (2) Since five amps plus the unknown current are coming in to the junction, and seven amps are leaving, KCL says that the total current in must equal the total current out, therefore the unknown current must be two amps in to the junction.

Let's take a look at a sample circuit, consisting of three 2000-ohm (2 kilo-ohm) resistors:

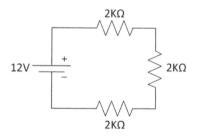

2KΩ

12V

2KΩ

2KΩ

There is only a single current path in the circuit, which travels through all three resistors. Instead of using three separate 2KΩ (2000Ω) resistors, you could replace the three resistors with one single resistor having an equivalent resistance. To find the equivalent resistance of any number of series resistors, just add up their individual resistances:

$$R_{eq} = R_1 + R_2 + R_3 + ...$$
$$R_{eq} = 2000\Omega + 2000\Omega + 2000\Omega$$
$$R_{eq} = 6000\Omega = 6K\Omega$$

Note that because there is only a single current path, the same current must flow through each of the resistors.

A simple and straightforward method for analyzing circuits involves creating a VIRP table for each circuit you encounter. Combining your knowledge of Ohm's Law, Kirchhoff's Current Law, Kirchhoff's Voltage Law, and equivalent resistance, you can use this table to solve for the details of any circuit.

A VIRP table describes the potential drop (V-voltage), current flow (I-current), resistance (R) and power dissipated (P-power) for each element in your circuit, as well as for the circuit as a whole. Let's use the circuit with the three 2000-ohm resistors as an example to demonstrate how a VIRP table is used. To create the VIRP table, first list the circuit elements, and total, in the rows of the table, then make columns for V, I, R, and P:

VIRP Table

	V	I	R	P
R_1				
R_2				
R_3				
Total				

Next, fill in the information in the table that is known. For example, you know the total voltage in the circuit (12V) provided by the battery, and you know the values for resistance for each of the individual resistors:

	V	I	R	P
R_1			2000Ω	
R_2			2000Ω	
R_3			2000Ω	
Total	12V			

Once the initial information has been filled in, you can also calculate the total resistance, or equivalent resistance, of the entire circuit. In this case, the equivalent resistance is 6000 ohms.

	V	I	R	P
R_1			2000Ω	
R_2			2000Ω	
R_3			2000Ω	
Total	12V		6000Ω	

Looking at the bottom (total) row of the table, both the voltage drop (V) and the resistance (R) are known. Using these two items, the total current flow in the circuit can be calculated using Ohm's Law.

$$I = \frac{V}{R} = \frac{12V}{6000\Omega} = 0.002\,A$$

The total power dissipated can also be calculated using any of the formulas for electrical power.

$$P = \frac{V^2}{R} = \frac{(12V)^2}{6000\Omega} = 0.024W$$

More information can now be completed in the VIRP table:

	V	I	R	P
R$_1$			2000Ω	
R$_2$			2000Ω	
R$_3$			2000Ω	
Total	12V	0.002A	6000Ω	0.024W

Because this is a series circuit, the total current has to be the same as the current through each individual element, so you can fill in the current through each of the individual resistors:

	V	I	R	P
R$_1$		0.002A	2000Ω	
R$_2$		0.002A	2000Ω	
R$_3$		0.002A	2000Ω	
Total	12V	0.002A	6000Ω	0.024W

Finally, for each element in the circuit, you now know the current flow and the resistance. Using this information, Ohm's Law can be applied to obtain the voltage drop (V=IR) across each resistor. Power can also be found for each element using P=I^{2}R to complete the table.

	V	I	R	P
R$_1$	4V	0.002A	2000Ω	0.008W
R$_2$	4V	0.002A	2000Ω	0.008W
R$_3$	4V	0.002A	2000Ω	0.008W
Total	12V	0.002A	6000Ω	0.024W

So what does this table really tell you now that it's completely filled out? You know the potential drop across each resistor (4V), the current through each resistor (2 mA), and the power dissipated by each resistor (8 mW). In addition, you know the total potential drop for the entire circuit is 12V, and the entire circuit dissipated 24 mW of power. Note that for a series circuit, the sum of the individual voltage drops across each element equal the total potential difference in the circuit, the current is the same throughout the circuit, and the resistances and power dissipated values add up to the total resistance and total power dissipated. These are summarized for you in the table below:

$$I = I_1 = I_2 = I_3 = ...$$
$$V = V_1 + V_2 + V_3 + ...$$
$$R_{eq} = R_1 + R_2 + R_3 + ...$$

11.28 Q: A 2.0-ohm resistor and a 4.0-ohm resistor are connected in series with a 12-volt battery. If the current through the 2.0-ohm resistor is 2.0 amperes, the current through the 4.0-ohm resistor is

(1) 1.0 A

(2) 2.0 A

(3) 3.0 A

(4) 4.0 A

11.28 A: (2) The current through a series circuit is the same everywhere, therefore the correct answer must be 2.0 amperes.

11.29 Q: In the circuit diagram below, two 4.0-ohm resistors are connected to a 16-volt battery as shown.

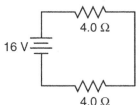

The rate at which electrical energy is expended in this circuit is

(1) 8.0 W

(2) 16 W

(3) 32 W

(4) 64 W

11.29 A: (3) 32W. Rate at which energy is expended is known as power.

	V	I	R	P
R$_1$	8V	2A	4Ω	16W
R$_2$	8V	2A	4Ω	16W
Total	16V	2A	8Ω	32W

11.30 Q: A 50-ohm resistor, an unknown resistor R, a 120-volt source, and an ammeter are connected in a complete circuit. The ammeter reads 0.50 ampere.

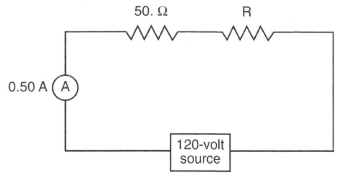

(A) Calculate the equivalent resistance of the circuit.

(B) Determine the resistance of resistor R.

(C) Calculate the power dissipated by the 50-ohm resistor.

11.30 A: (A) $R_{eq} = 240\Omega$ (B) R= 190Ω (C) $P_{50\Omega\ resistor} = 12.5W$

	V	I	R	P
R₁	25V	0.50A	50Ω	12.5W
R₂	95V	0.50A	190Ω	47.5W
Total	120V	0.50A	240Ω	60W

11.31 Q: What must be inserted between points A and B to establish a steady electric current in the incomplete circuit represented in the diagram?

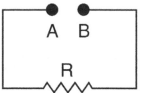

(1) switch

(2) voltmeter

(3) magnetic field source

(4) source of potential difference

11.31 A: (4) a source of potential difference is required to drive current.

11.32 Q: In the circuit represented by the diagram, what is the reading of voltmeter V?

(1) 20 V

(2) 2.0 V

(3) 30 V

(4) 40 V

11.32 A: (4) Voltmeter reads potential difference across R_1 which is 40 V.

	V	**I**	**R**	**P**
R_1	40V	2A	20Ω	80W
R_2	20V	2A	10Ω	40W
Total	60V	2A	30Ω	120W

Parallel Circuits

Another basic circuit type is the **parallel circuit**, in which there is more than one current path. To analyze resistors in a series circuit, you found an equivalent resistance. You'll follow the same strategy in analyzing resistors in parallel.

Let's examine a circuit made of the same components used in the exploration of series circuits, but now connect the components so as to provide multiple current paths, creating a parallel circuit.

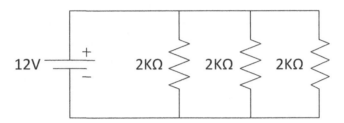

Notice that in this circuit, electricity can follow one of three different paths through each of the resistors. In many ways, this is similar to a river branching into three different smaller rivers. Each resistor, then, causes a potential drop (analogous to a waterfall), then the three rivers recombine before heading back to the battery, which you can think of like a pump, raising the river to a higher potential before sending it back on its looping path. Or you can think of it as students rushing out of a classroom. The more doors in the room, the less resistance there is to exiting!

You can find the equivalent resistance of resistors in parallel using the formula:

$$\frac{1}{R_{eq}} = \frac{1}{R_1} + \frac{1}{R_2} + \frac{1}{R_3} + \dots$$

Take care in using this equation, as it's easy to make errors in performing your calculations. For only two resistors, this simplifies to:

$$R_{eq} = \frac{R_1 R_2}{R_1 + R_2}$$

Let's find the equivalent resistance for the sample circuit.

$$\frac{1}{R_{eq}} = \frac{1}{R_1} + \frac{1}{R_2} + \frac{1}{R_3} + ...$$

$$\frac{1}{R_{eq}} = \frac{1}{2000\Omega} + \frac{1}{2000\Omega} + \frac{1}{2000\Omega}$$

$$\frac{1}{R_{eq}} = 0.0015\,\text{\Large/}_\Omega$$

$$R_{eq} = \frac{1}{0.0015\,\text{\Large/}_\Omega} = 667\Omega$$

A VIRP table can again be used to analyze the circuit, beginning by filling in what is known directly from the circuit diagram.

VIRP Table

	V	I	R	P
R$_1$			2000Ω	
R$_2$			2000Ω	
R$_3$			2000Ω	
Total	12V			

You can also see from the circuit diagram that the potential drop across each resistor must be 12V, since the ends of each resistor are held at a 12-volt difference by the battery

	V	I	R	P
R$_1$	12V		2000Ω	
R$_2$	12V		2000Ω	
R$_3$	12V		2000Ω	
Total	12V			

Next, you can use Ohm's Law to fill in the current through each of the individual resistors since you know the voltage drop across each resistor (I=V/R) to find I=0.006A.

	V	I	R	P
R$_1$	12V	0.006A	2000Ω	
R$_2$	12V	0.006A	2000Ω	
R$_3$	12V	0.006A	2000Ω	
Total	12V			

Using Kirchhoff's Current Law, you can see that if 0.006A flows through each of the resistors, these currents all come together to form a total current of 0.018A.

	V	I	R	P
R$_1$	12V	0.006A	2000Ω	
R$_2$	12V	0.006A	2000Ω	
R$_3$	12V	0.006A	2000Ω	
Total	12V	**0.018A**		

Because each of the three resistors has the same resistance, it only makes sense that the current would be split evenly between them. You can confirm the earlier calculation of equivalent resistance by calculating the total resistance of the circuit using Ohm's Law: R=V/I=(12V/0.018A)=667Ω.

	V	I	R	P
R$_1$	12V	0.006A	2000Ω	
R$_2$	12V	0.006A	2000Ω	
R$_3$	12V	0.006A	2000Ω	
Total	12V	0.018A	**667Ω**	

Finally, you can complete the VIRP table using any of the three applicable equations for power dissipation to find:

	V	I	R	P
R$_1$	12V	0.006A	2000Ω	**0.072W**
R$_2$	12V	0.006A	2000Ω	**0.072W**
R$_3$	12V	0.006A	2000Ω	**0.072W**
Total	12V	0.018A	667Ω	**0.216W**

Note that for resistors in parallel, the equivalent resistance is always less than the resistance of any of the individual resistors. The potential difference across each of the resistors in parallel is the same, and the current through each of the resistors adds up to the total current. This is summarized for you in the following table:

$$I = I_1 + I_2 + I_3 + ...$$
$$V = V_1 = V_2 = V_3 = ...$$
$$\frac{1}{R_{eq}} = \frac{1}{R_1} + \frac{1}{R_2} + \frac{1}{R_3} + ...$$

11.33 Q: A 15-ohm resistor, R_1, and a 30-ohm resistor, R_2, are to be connected in parallel between points A and B in a circuit containing a 90-volt battery.

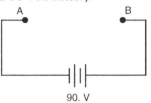

90. V

(A) Complete the diagram to show the two resistors connected in parallel between points A and B.

(B) Determine the potential difference across resistor R_1.

(C) Calculate the current in resistor R_1.

11.33 A: (A)

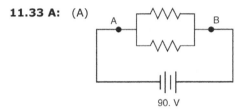

90. V

(B) Potential difference across R_1 is 90V.

(C) Current through resistor R_1 is 6A.

	V	I	R	P
R_1	90V	6A	15Ω	540W
R_2	90V	3A	30Ω	270W
Total	90V	9A	10Ω	810W

11.34 Q: Draw a diagram of an operating circuit that includes: a battery as a source of potential difference, two resistors in parallel with each other, and an ammeter that reads the total current in the circuit.

11.34 A:

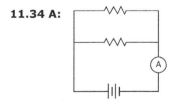

11.35 Q: Three identical lamps are connected in parallel with each other. If the resistance of each lamp is X ohms, what is the equivalent resistance of this parallel combination?

(1) X Ω

(2) X/3 Ω

(3) 3X Ω

(4) 3/X Ω

11.35 A: (2) X/3 Ω

$$\frac{1}{R_{eq}} = \frac{1}{R_1} + \frac{1}{R_2} + \frac{1}{R_3} + ...$$

$$\frac{1}{R_{eq}} = \frac{1}{X} + \frac{1}{X} + \frac{1}{X}$$

$$\frac{1}{R_{eq}} = \frac{3}{X}$$

$$R_{eq} = \frac{X}{3}$$

11.36 Q: Three resistors, 4 ohms, 6 ohms, and 8 ohms, are connected in parallel in an electric circuit. The equivalent resistance of the circuit is

(1) less than 4 Ω

(2) between 4 Ω and 8 Ω

(3) between 10 Ω and 18 Ω

(4) 18 Ω

11.36 A: (1) the equivalent resistance of resistors in parallel is always less than the value of the smallest resistor.

11.37 Q: A 3-ohm resistor, an unknown resistor, R, and two ammeters, A_1 and A_2, are connected as shown with a 12-volt source. Ammeter A_2 reads a current of 5 amperes.

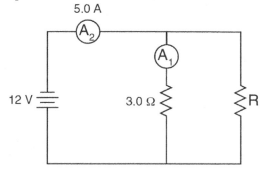

(A) Determine the equivalent resistance of the circuit.

(B) Calculate the current measured by ammeter A_1.

(C) Calculate the resistance of the unknown resistor, R.

11.37 A: (A) 2.4Ω (B) 4A (C) 12Ω

	V	I	R	P
R_1	12V	4A	3Ω	48W
R_2	12V	1A	12Ω	12W
Total	12V	5A	2.4Ω	60W

11.38 Q: The diagram below represents an electric circuit consisting of four resistors and a 12-volt battery.

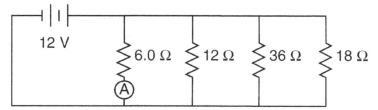

(A) What is the current measured by ammeter A?

(B) What is the equivalent resistance of this circuit?

(C) How much power is dissipated in the 36-ohm resistor?

11.38 A: (A) 2A (B) 3Ω (C) 4W

	V	I	R	P
R_1	12V	2A	6Ω	24W
R_2	12V	1A	12Ω	12W
R_3	12V	0.33A	36Ω	4W
R_4	12V	0.67A	18Ω	8W
Total	12V	4A	3Ω	48W

11.39 Q: A 20-ohm resistor and a 30-ohm resistor are connected in parallel to a 12-volt battery as shown. An ammeter is connected as shown.

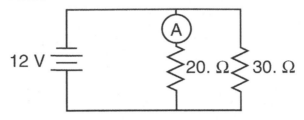

(A) What is the equivalent resistance of the circuit?

(B) What is the current reading of the ammeter?

(C) What is the power of the 30-ohm resistor?

11.39 A: (A) 12Ω   (B) 0.6A   (C) 4.8W

	V	I	R	P
R₁	12V	0.6A	20Ω	7.2W
R₂	12V	0.4A	30Ω	4.8W
Total	12V	1A	12Ω	12W

11.40 Q: In the circuit diagram shown below, ammeter A_1 reads 10 amperes.

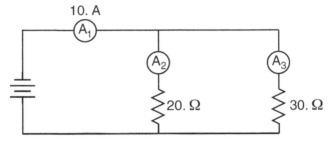

What is the reading of ammeter A_2?

(1) 6 A

(2) 10 A

(3) 20 A

(4) 4 A

11.40 A: (1) 6 A

	V	I	R	P
R₁	120V	6A	20Ω	720W
R₂	120V	4A	30Ω	480W
Total	120V	10A	12Ω	1200W

Combination Series-Parallel Circuits

A circuit doesn't have to be completely serial or parallel. In fact, most circuits actually have elements of both types. Analyzing these circuits can be accomplished using the fundamentals you learned in analyzing series and parallel circuits separately and applying them in a logical sequence.

First, look for portions of the circuit that have parallel elements. Since the voltage across the parallel elements must be the same, replace the parallel resistors with an equivalent single resistor in series and draw a new schematic. Now you can analyze your equivalent series circuit with a VIRP table. Once your table is complete, work back to your original circuit using KCL and KVL until you know the current, voltage, and resistance of each individual element in your circuit.

11.41 Q: Find the current through R_2 in the circuit below.

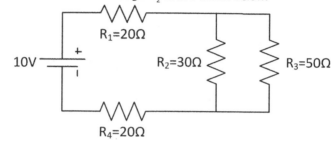

11.41 A: First, find the equivalent resistance for R_2 and R_3 in parallel.

$$R_{eq_{23}} = \frac{R_2 R_3}{R_2 + R_3} = \frac{(30\Omega)(50\Omega)}{30\Omega + 50\Omega} = 19\Omega$$

Next, re-draw the circuit schematic as an equivalent series circuit.

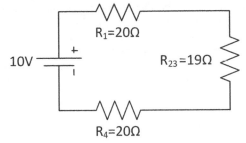

Now, you can use your VIRP table to analyze the circuit.

	V	I	R	P
R_1	3.39V	0.169A	20Ω	0.57W
R_{23}	3.22V	0.169A	19Ω	0.54W
R_4	3.39V	0.169A	20Ω	0.57W
Total	10V	0.169A	59Ω	1.69W

The voltage drop across R_2 and R_3 must therefore be 3.22 volts. From here, you can apply Ohm's Law to find the current through R_2:

$$I_2 = \frac{V_2}{R_2} = \frac{3.22V}{30\Omega} = 0.107\,A$$

Chapter 12: Magnetism

"Magnetism, as you recall from physics class, is a powerful force that causes certain items to be attracted to refrigerators."

— *Dave Barry*

Objectives

1. Explain that magnetism is caused by moving electrical charges.
2. Describe the magnetic poles and interactions between magnets.
3. Draw magnetic field lines.
4. Describe the factors affecting an induced potential difference due to magnetic field lines interacting with moving charges.
5. Describe the three right hand rules for magnetism.
6. Calculate the force exerted on a charge moving through a magnetic field.

Magnetism is closely related to electricity. In essence, **magnetism** is a force caused by moving charges. In the case of permanent magnets, the moving charges are the orbits of electrons spinning around nuclei. In very basic terms, strong permanent magnets have many atoms with electrons spinning in the same direction. Non-magnets have more random arrangements of electrons spinning around the nucleus. For electromagnets, the current itself provides the moving charges. In all cases, magnetic fields can be used to describe the forces due to magnets.

12.01 Q: Which type of field is present near a moving electric charge?
(1) an electric field, only
(2) a magnetic field, only
(3) both an electric field and a magnetic field
(4) neither an electric field nor a magnetic field

12.01 A: (3) An electric field is present due to the electric charge, and a magnetic field is present because the charge is in motion.

Magnetic Fields

Magnets are polarized, meaning every magnet has two opposite ends. The end of a magnet that points toward the geographic north pole of the Earth is called the north pole of the magnet, while the opposite end, for obvious reasons, is called the magnet's south pole. Every magnet has both a north and a south pole. There are no single isolated magnetic poles, or monopoles. If you split a magnet in half, each half of the original magnet exhibits both a north and a south pole, giving you two magnets. Physicists continue to search both physically and theoretically, but to date, no one has ever observed a north pole without a south pole, or a south pole without a north pole.

You used electric field lines to help visualize what would happen to a positive charge placed in an electric field. In order to visualize a magnetic field, you can draw magnetic field lines (also known as magnetic flux lines) which show the direction the north pole of a magnet would tend to point if placed in the field. Magnetic field lines are drawn as closed loops, starting from the north pole of a magnet and continuing to the south pole of a magnet. Inside the magnet itself, the field lines run from the south pole to the north pole. The magnetic field is strongest in areas of greatest density of magnetic field lines, or areas of the greatest magnetic flux density. Magnetic field strength (B) is measured in units known as Tesla (T).

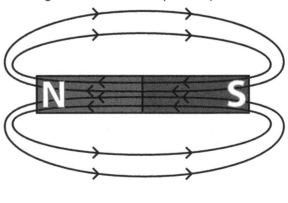

Much like electrical charges, like poles exert a repelling force on each other, while opposite poles exert an attractive force on each other. Materials can be classified as magnets, magnet attractables (materials which aren't magnets themselves but can be attracted by magnets), and non-attractables.

12.02 Q: The diagram below shows the lines of magnetic force between two north magnetic poles. At which point is the magnetic field strength greatest?

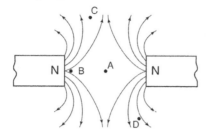

12.02 A: (B) has the greatest magnetic field strength because it is located at the highest density of magnetic field lines.

12.03 Q: The diagram below represents a 0.5-kilogram bar magnet and a 0.7-kilogram bar magnet with a distance of 0.2 meter between their centers.

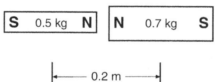

Which statement best describes the forces between the bar magnets?

(1) Gravitational force and magnetic force are both repulsive.

(2) Gravitational force is repulsive and magnetic force is attractive.

(3) Gravitational force is attractive and magnetic force is repulsive.

(4) Gravitational force and magnetic force are both attractive.

12.03 A: (3) Gravity always attracts and the north poles repel each other.

12.04 Q: A student is given two pieces of iron and told to determine if one or both of the pieces are magnets. First, the student touches an end of one piece to one end of the other. The two pieces of iron attract. Next, the student reverses one of the pieces and again touches the ends together. The two pieces attract again. What does the student definitely know about the initial magnetic properties of the two pieces of iron?

12.04 A: At least one of the pieces of iron is a magnet, but we cannot state with certainty that both are magnets.

12.05 Q: Draw a minimum of four field lines to show the magnitude and direction of the magnetic field in the region surrounding a bar magnet.

12.05 A:

12.06 Q: When two ring magnets are placed on a pencil, magnet A remains suspended above magnet B, as shown below.

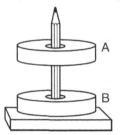

Which statement describes the gravitational force and the magnetic force acting on magnet A due to magnet B?

(1) The gravitational force is attractive and the magnetic force is repulsive.

(2) The gravitational force is repulsive and the magnetic force is attractive.

(3) Both the gravitational force and the magnetic force are attractive.

(4) Both the gravitational force and the magnetic force are repulsive.

12.06 A: (1) Gravity can only attract, and because magnet A is suspended above magnet B, the magnetic force must be repulsive.

The Compass

Because the Earth exerts a force on magnets (which, when used to tell direction, we call a compass), you can conclude that the Earth is a giant magnet. If the north pole of a magnet is attracted to the geographic north pole of the Earth, and opposite poles attract, then it stands to reason that the geographic north pole of the Earth is actually a magnetic south pole! Compasses always line up with the net magnetic field.

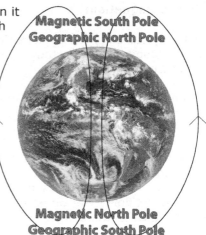

Magnetic South Pole
Geographic North Pole

In truth, the magnetic north and south pole of the Earth are constantly moving. The current rate of change of the magnetic north pole is thought to be more than 20 kilometers per year, and it is believed that the magnetic north pole has shifted more than 1000 kilometers since it was first reached by explorer Sir John Ross in 1831!

Magnetic North Pole
Geographic South Pole

12.07 Q: The diagram below represents the magnetic field near point P.

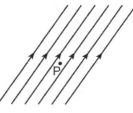

If a compass is placed at point P in the same plane as the magnetic field, which arrow represents the direction the north end of the compass needle will point?

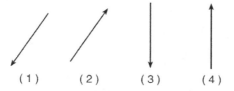

(1)   (2)   (3)   (4)

12.07 A: (2) Compass needles line up with the magnetic field.

12.08 Q: The diagram below shows a bar magnet.

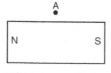

A

N   S

Which way will the needle of a compass placed at A point?

(1) up   (3) right
(2) down   (4) left

12.08 A: (3) since a compass lines up with the magnetic field.

Electromagnetism

In 1820, Danish physicist Hans Christian Oersted found that a current running through a wire created a magnetic field, kicking off the modern study
of **electromagnetism**.

Moving electric charges create magnetic fields. You can test this by placing
a compass near a current-carrying wire. The compass will line up with the
induced magnetic field.

To determine the direction of the electrically-induced
magnetic field, use the first right hand rule (RHR) by
pointing your right-hand thumb in the direction of positive current flow. The curve of your fingers then shows
the direction of the magnetic field around a wire.

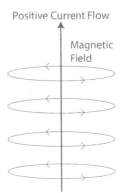

You can obtain an even stronger magnetic field by wrap-
ping a coil of wire in a series of loops known as a solenoid and flowing current through the wire. This is known
as an electromagnet. You can make the magnetic field
from the electromagnet even stronger by placing a piece
of iron inside the coils of wire. The second right hand
rule tells you the direction of the magnetic field due to
an electromagnet. Wrap your fingers around the solenoid in the direction
of positive current flow. Your thumb will point toward the north end of the
induced magnetic field.

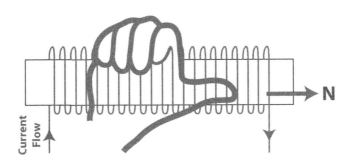

Not only do moving charges create magnetic fields, but relative motion between charges and a magnetic field can produce a force. The magnitude of
the force (F_B) on a charge (q) moving through a magnetic field (B) with a
velocity (v) is given by:

$$F_B = qvB \sin \theta$$

In this equation, θ is the angle between the velocity vector and the direction of the magnetic field. If the velocity of the charged particle is perpendicular to the magnetic field, sin θ = sin 90° = 1, and the force can be calculated as simply F_B=qvB.

Because force is a vector, it has a direction as well. This direction can be determined using the third right hand rule. Point the fingers of your right hand in the direction of a positive particle's velocity (if the moving charge is negative, point the fingers of your right hand in a direction opposite the particle's velocity). Then, curl your fingers inward 90° in the direction the magnetic field points. Your thumb will point in the direction of the force on the charged particle.

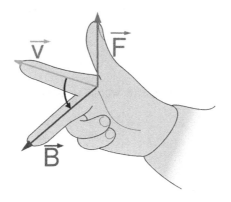

Illustration Courtesy of David Crochet

12.09 Q: An electron moves at 2.0×10⁶ meters per second perpendicular to a magnetic field having a flux density of 2.0 teslas. What is the magnitude of the magnetic force on the electron?

(1) 1.0×10⁻⁶ N

(2) 6.4×10⁻¹³ N

(3) 3.6×10⁻²⁴ N

(4) 4.0×10⁶ N

12.09 A: $F_B = qvB \sin\theta \rightarrow$

$F_B = (-1.6\times10^{-19}C)(2\times10^6 \, ^m\!/_s)(2T)\sin 90° = -6.4\times10^{-13}N$

12.10 Q: A particle with a charge of 6.4×10⁻¹⁹ C experiences a force of 2×10⁻¹² N as it travels through a 3 tesla magnetic field at an angle of 30° to the field. What is the particle's velocity?

12.10 A: $F_B = qvB\sin\theta \rightarrow v = \dfrac{F_B}{qB\sin\theta} \rightarrow$

$$v = \frac{2\times10^{-12}\,N}{(6.4\times10^{-19}\,C)(3T)(\sin 30°)} = 2.08\times10^{6}\,{}^{m}\!/\!_{s}$$

12.11 Q: The air core of the electromagnet is replaced with an iron core. Compared to the strength of the magnetic field in the air core, the strength of the magnetic field in the iron core is

(1) less

(2) greater

(3) the same

12.11 A: (2) An iron core placed within an electromagnet strengthens the magnetic field.

12.12 Q: The diagram below shows a proton moving with velocity v about to enter a uniform magnetic field directed into the page. As the proton moves in the magnetic field, the magnitude of the magnetic force on the proton is F.

If the proton were replaced by an alpha particle (charge +2e) under the same conditions, the magnitude of the magnetic force on the alpha particle would be

(1) F

(2) 2F

(3) F/2

(4) 4F

12.12 A: (2) 2F. Because charge is doubled, magnetic force also doubles.

When relative motion between a conductor and a magnetic field creates a force on the charges in the conductor, a potential difference is induced in the conductor. The conductor must cut across the magnetic field lines to produce a potential difference, and larger potential differences are created when the conductor cuts across stronger magnetic fields, or moves more quickly through the magnetic field.

This phenomenon is what allows you to create usable, controllable electrical energy. Kinetic energy in the form of wind, water, steam, etc. is used to spin a coil of wire through a magnetic field, inducing a potential difference, which is transferred by the electric company to end users. This basic energy transformation is the underlying principle behind hydroelectric, nuclear, fossil fuel, and wind-powered electrical generators!

12.13 Q: The diagram below shows a wire moving to the right at speed v through a uniform magnetic field that is directed into the page. As the speed of the wire is increased, the induced potential difference will

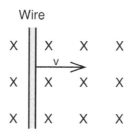

Magnetic field directed into page

(1) decrease
(2) increase
(3) remain the same

12.13 A: (2) the induced potential difference will increase as the speed of the wire is increased.

12.14 Q: The diagram below represents a wire conductor, RS, positioned perpendicular to a uniform magnetic field directed into the page.

		R			
X	X		X	X	Magnetic
X	X		X	X	field
X	X		X	X	directed
X	X		X	X	into the page
		S			

Describe the direction in which the wire could be moved to produce the maximum potential difference across its ends, R and S.

12.14 A: The wire could be moved to produce the maximum potential difference across its ends, R and S, by moving it horizontally (right to left or left to right).

Chapter 12: Magnetism

Chapter 13: Microelectronics

"The complexity for minimum component costs has increased at a rate of roughly a factor of two per year...

Certainly over the short term this rate can be expected to continue, if not to increase."

— Gordon Moore, 1965

Objectives

1. Describe the characteristics of a semiconductor.
2. Qualitatively describe the operation of diodes and FET transistors.
3. Develop vocabulary needed to converse in the semiconductor and nanotechnology industries.
4. Explain how transistors may be connected to create simple logic gates.
5. Utilize basic logic gates to build simple digital logic devices such as a half adder.
6. Describe the basic process steps in integrated circuit fabrication.

Integrated Circuits

In 1947 at AT&T Bell Laboratories in Murray Hills, N.J., three scientists, John Bardeen, Walter Brattain, and William Shockley, invented the bipolar junction transistor, eventually winning the Nobel Prize for their work.

Eleven years later, in 1958, Jack Kilby of Texas Instruments created the first integrated circuit, which is comprised of more than one device built on the same substrate. Bob Noyce of Fairchild Semiconductor followed this up with a patent on an integrated circuit based on a planar process. In 1965, Gordon Moore, also of Fairchild, predicted that the number of devices on a chip would double every two years by making them smaller. Today devices are built with billions of transistors on a single chip!

You can find integrated circuits everywhere. From computers, cell phones, televisions and radios to watches, cameras, refrigerators, ovens, and automobiles, integrated circuits have invaded almost every aspect of modern day society, and play roles in obvious arenas such as entertainment and productivity, but also play an important part in safety, research, education, communication, agriculture, transportation, medicine, and more!

Silicon

Semiconductors come in two types, elemental and compound. The elemental semiconductors come from column IV in the periodic table, and are silicon and germanium. Silicon is used more commonly than germanium due to its ability to withstand high temperatures and its ability to oxidize.

Periodic Table of the Elements

I																	VIII
1 H	II											III	IV	V	VI	VII	2 He
3 Li	4 Be											5 B	6 C	7 N	8 O	9 F	10 Ne
11 Na	12 Mg											13 Al	14 Si	15 P	16 S	17 Cl	18 Ar
19 K	20 Ca	21 Sc	22 Ti	23 V	24 Cr	25 Mn	26 Fe	27 Co	28 Ni	29 Cu	30 Zn	31 Ga	32 Ge	33 As	34 Se	35 Br	36 Kr
37 Rb	38 Sr	39 Y	40 Zr	41 Nb	42 Mo	43 Tc	44 Ru	45 Rh	46 Pd	47 Ag	48 Cd	49 In	50 Sn	51 Sb	52 Te	53 I	54 Xe
55 Cs	56 Ba		72 Hf	73 Ta	74 W	75 Re	76 Os	77 Ir	78 Pt	79 Au	80 Hg	81 Tl	82 Pb	83 Bi	84 Po	85 At	86 Rn
87 Fr	88 Ra		104 Rf	105 Db	106 Sg	107 Bh	108 Hs	109 Mt	110 Ds	111 Rg	112 Cn	113 Uut	114 Uuq	115 Uup	116 Uuh	117 Uus	118 Uuo

Lanthanoids	57 La	58 Ce	59 Pr	60 Nd	61 Pm	62 Sm	63 Eu	64 Gd	65 Tb	66 Dy	67 Ho	68 Er	69 Tm	70 Yb	71 Lu
Actanoids	89 Ac	90 Th	91 Pa	92 U	93 Np	94 Pu	95 Am	96 Cm	97 Bk	98 Cf	99 Es	100 Fm	101 Md	102 No	103 Lr

Semiconductors can also be made from compounds such as gallium arsenide, indium phosphide, and cadmium telluride. These compounds may be faster and more resistant to radiation, yet silicon is still far and away the most popular semiconductor for the production of integrated circuits. Silicon is cheaper than its competitors. Silicon is easier to process than its competitors. And silicon is the 2nd most abundant element on earth (behind oxygen).

Silicon has an atomic number of 14, and has four valence electrons. This indicates that silicon forms primarily covalent bonds, and when found in its pure crystalline form, each silicon atom shares an electron with its four nearest neighbor silicon atoms. The shared electrons are very tightly bound to the atoms, leaving very few electrons free to wander throughout the solid. This makes pure single crystal silicon a fairly good insulator.

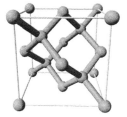

In single crystal form, silicon forms a diamond lattice, with a unit cell having eight full silicon atoms. The diamond lattice is actually two inter-penetrating face-centered-cubic (FCC) lattices, displaced one quarter of the distance along the major diagonal. As you can imagine, this makes for a very highly ordered, densely packed material. Each side of the lattice itself is 5.43×10^{-10} m.

If you replace one of the silicon atoms in the lattice with an impurity atom known as a **dopant**, you can modify the electrical properties of the structure. By replacing a silicon atom with an atom that has five valence electrons, from column five of the periodic table, such as phosphorus, arsenic, or antimony, four of the electrons would be required for covalent bonding, leaving an extra electron "free" to wander the lattice. These dopant atoms are known as **donors**, because they "donate" their extra electron to the lattice, effectively increasing the number of negative charge carriers in the solid, which increases the material's conductivity (and decreases its resistivity).

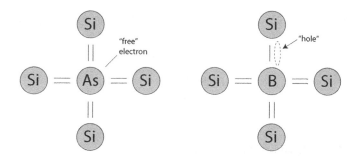

In similar fashion, if you were to replace one of the silicon atoms in the lattice with an atom that has three valence electrons, from column three of the period table, such as boron, indium, gallium, or aluminum, three of the electrons would be used up in covalent bonding, leaving a **hole** where a bond was previously. Electrons can jump in and fill that hole, leaving a hole elsewhere, in effect allowing the hole to move, therefore you have created a positive charge carrier, which also increases the material's conductivity and

Chapter 13: Microelectronics

decreases its resistivity. These dopant atoms are known as **acceptors**, because they accept an electron from the lattice.

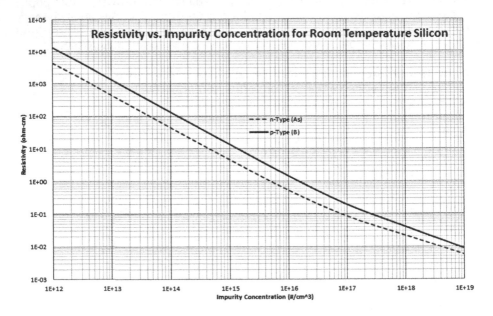

When silicon has been doped with more donors than acceptors, you have an excess of free electrons, or negative charges. This is called **n-type silicon**. When silicon has been doped with more acceptors than donors, providing an excess of free holes, or positive charges, it is called **p-type silicon**. Donor atoms are therefore also known as n-type dopants, and acceptor atoms are also known as p-type dopants. Note that in the case of both n-type and p-type dopants, the material as a whole remains neutral because the extra charge carriers are offset by the charge of the dopant nuclei.

13.1 Q: Magnetic-card door locks utilize many electronic components on one small piece of semiconductor material. This combination of components on a single chip is called

(1) a transistor

(2) an integrated circuit

(3) a printed circuit board

(4) a diode

13.1 A: (2) an integrated circuit

13.2 Q: Current in a semiconductor is caused by the movement of

(1) electrons only　　　　(2) holes only

(3) isotopes　　　　(4) both electrons and holes

13.2 A: (4) both electrons and holes

13.3 Q: The diagram below shows a circuit with a battery applying a potential difference across a p-type semiconductor.

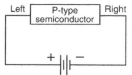

The majority charge carriers in the semiconductor are

(1) negative electrons moving to the right

(2) negative electrons moving to the left

(3) positive holes moving to the right

(4) positive holes moving to the left

13.3 A: (3) positive holes moving to the right.

13.4 Q: Donor materials are added to semiconductors so that the number of available electrons will

(1) decrease

(2) increase

(3) remain the same

13.4 A: (2) Donors increase the number of available electrons.

13.5 Q: An impurity that is added to a semiconductor in order to provide holes is classified as a

(1) donor

(2) receptor

(3) acceptor

(4) bias

13.5 A: (3) acceptors accept an electron, leaving a hole.

13.6 Q: Investigate what has happened to Bob Noyce and Gordon Moore since 1965.

P-N Junctions

When p-type silicon is placed in close contact with n-type silicon (in what is known as a metallurgical junction), a **p-n junction** is formed. P-N junctions are electrical devices more commonly known as **diodes**. Through careful analysis of the junction, you can build a qualitative understanding of how these devices operate.

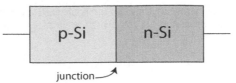

junction

Free electrons on the n-side of the junction diffuse across the junction into the p-side, leaving behind positively ionized donor atoms (N_D^+) as they slide into the vacancies that are previously holes. This means the electrons have disappeared in the p-type silicon and leaves the n-type silicon region close to the junction with a net positive charge.

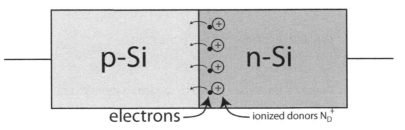

electrons — ionized donors N_D^+

In similar fashion, holes on the p-side of the junction diffuse across the junction into the n-side, leaving behind negatively ionized acceptor atoms (N_A^-) The region close to the junction in the p-type silicon picks up a net negative charge. The holes disappear in the n-type silicon because of all the free electrons.

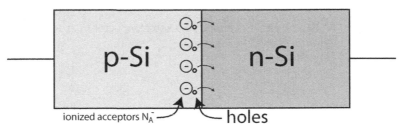

ionized acceptors N_A^- — holes

Because the free charges near the junction have crossed the junction and been disappeared, there is a deficiency of free charges near the junction. This region is known as the **depletion region**, because it is depleted of free carriers. You may also hear it referred to as the **space charge region**, because it is a region of positive donor charges and negative acceptor charges separated by space.

The depletion region on the p-side of the junction is negatively charged because the dopant acceptor atoms picked up electrons, and the depletion region on the n-side of the junction is positively charged because the dopant donor atoms lost their electrons. Note that the ionized dopant atoms are not free to move, unlike the electrons and holes, because the dopants are held in place in the crystal lattice.

Chapter 13: Microelectronics

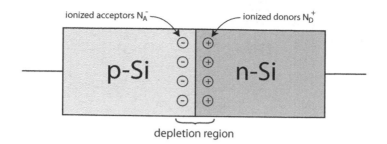

ionized acceptors N_A^- — ionized donors N_D^+

depletion region

The fixed charges in the depletion region set up a small electric field pointing from the positive side to the negative side. This is known as the built-in electric field ($\mathcal{E}_{bi}$).

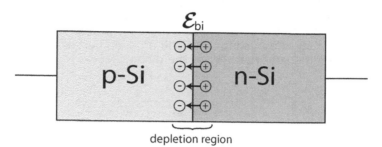

$\mathcal{E}_{bi}$

depletion region

In the bulk of the p-n junction, away from the depletion region, the abundance of free carriers capable of transporting charge makes the regions good conductors. In the small depletion region near the junction, however, the absence of free carriers makes the depletion region a good insulator. Further, positive charge carriers (holes) from the p-side bulk that wander near the depletion region get pushed back into the p-side bulk by the built-in electric field. Likewise, negative charge carriers (electrons) from the n-side bulk wandering near the depletion region are pushed back into the n-side bulk. The net effect is that current cannot flow from the p-side to the n-side of the diode. Carriers that try to diffuse will be returned by the electric field.

When a potential difference of approximately 0.7 volts or greater is applied to the circuit, however, with the positive side of the applied voltage connected to the p-side of the diode, the applied voltage counteracts the built-in field in the depletion region, and charge carriers can again flow across the junction by diffusion, allowing positive current flow from the p-side to the n-side of the diode. This is known as applying a **forward bias** to the diode.

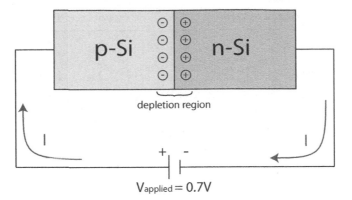

Forward Bias - Current Flow p to n

If a potential difference (voltage) is applied in the opposite direction, however, with the positive side of the applied voltage attached to the n-side of the diode, the applied voltage enhances the built-in field in the diode, preventing current flow and making the diode an even better insulator. This is known as applying a **reverse bias** to the diode.

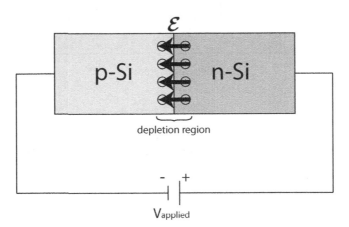

Reverse Bias - No Current Flow

13.7 Q: An n-type semiconductor and a p-type semiconductor are joined to form a diode. Compared to the total number of electrons in the semiconductors before joining, the number of electrons in the diode is

(1) fewer

(2) greater

(3) the same

13.7 A: (3) the same due to the law of conservation of charge.

13.8 Q: Using the diagram below, answer the following questions

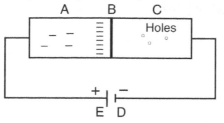

A. In the diagram, C represents the
(1) n-type silicon   (2) p-type silicon
(3) anode   (4) diode

B. The p-n junction in the diagram is biased
(1) reverse   (2) forward
(3) A to E   (4) C to D

13.8 A: A. (2) p-type silicon (you can tell because of all the holes!)
         B. (1) the p-n junction is reverse biased.

13.9 Q: Compared to the current flow when a forward bias is applied to a p-n junction, the current flow when a reverse bias is applied to a p-n junction is
(1) less
(2) greater
(3) the same

13.9 A: (1) the current through a reverse-biased p-n junction is minimal.

13.10 Q: In the p-n junction region of an operating diode, an electric field barrier is produced by free electrons in the
(1) n-type material crossing into the p-type material
(2) n-type material going away from the p-type material
(3) p-type material crossing into the n-type material
(4) p-type material going away from the n-type material

13.10 A: (1) electrons in the n-type silicon near the junction move from the n-type material into the p-type material.

Putting this all together, you can make a graph of the current through a diode as a function of the applied potential difference V_{app}. This is known as an **I-V curve**, where I is current and V is the applied potential difference.

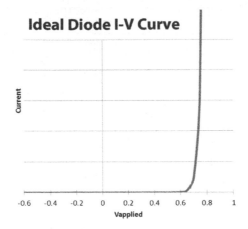

Ideal Diode I-V Curve

As you can see from the diode's I-V curve, the diode acts like a one-way gate for current, allowing current to flow through in one direction only (from p to n). For this reason, the circuit symbol for a diode indicates the direction current can flow from p to n with an arrow, and shows a line blocking current from n to p.

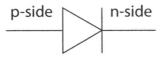

p-side   n-side

current flows this way only

When placed in a circuit, a diode acts as an insulator, or open, until the diode is given a forward bias of approximately 0.7 volts, at which point it acts in a fashion similar to a very low resistance wire. Circuit analysis of a diode, therefore, can be quite straightforward.

13.11 Q: Find the current flowing in the circuit below. Assume the diode turns on at exactly 0.7 volts of forward bias.

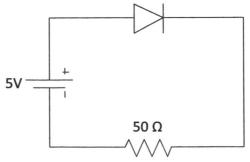

13.11 A: Because the voltage source provides more than 5V of forward bias, you can assume the diode is turned on. Therefore, the voltage drop across the diode must be 0.7 volts, leaving 4.3 volts to drop across the resistor. You can find the current through the resistor (and therefore through the entire circuit since the circuit is in series) using Ohm's Law.

$$I = \frac{V}{R} = \frac{4.3V}{50\Omega} = 0.086\,A$$

13.12 Q: Find the current flow through the circuit below. Assume the diodes turn on at exactly 0.7 volts of forward bias.

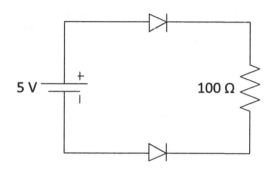

13.12 A: The diodes in this circuit are placed in opposing directions, blocking all current flow. Therefore, the current flow through this circuit is zero.

Transistors

What happens when two diodes are placed back-to-back in a p-n-p configuration, as shown in the diagram below?

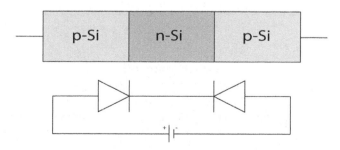

Current cannot flow. Because a diode acts like a one-way gate for current, whichever direction current comes from, one of the diodes is reverse biased, cutting off current flow through the circuit. But, what if you could selectively modify the center region?

By placing a second, negative potential near the n-type silicon region, electrons near the negative potential will be repelled, creating another depletion region in the n-type silicon.

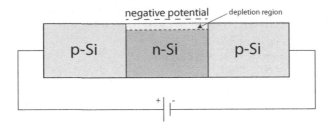

As the depletion region grows, the bias across the p-n junction on the left approaches 0.7 V and the holes in the p-type silicon enter the depleted n-region. The holes stay near the surface and form a channel across the n-Si to the p-Si on the right. This is known as a p-channel inside the n-type silicon, as you have "inverted" the silicon.

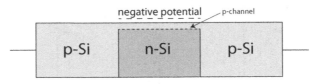

You now have a complete p-type silicon path with ample charge carriers, therefore current can flow through the device. The p-n-p device is now acting like a controllable gateway for current, and the negative potential is the switch (or gate) to turn on and off the current flow from what is known as the source side of the device to the drain side of the device, as shown below.

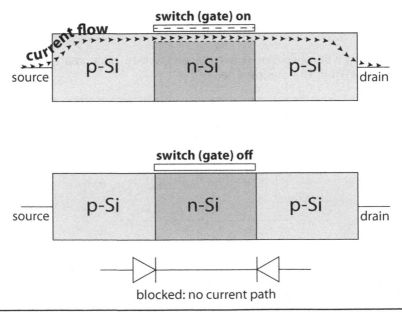

This p-n-p device, with three contacts (source of current, drain of current, and gate switch), is known as a **field effect transistor**, or FET. The circuit schematic symbol for a pnp FET indicates the source, gate, and drain of the device, as shown below.

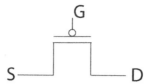

Note that there are a number of other types and modifications of FETs, including devices made out of n-p-n junctions. For purposes of simplicity, this book will focus solely on the p-n-p device described. Readers are referred to more detailed texts on microelectronic technology and digital logic for a more thorough and complete background.

Digital Logic

Now that you know how a single transistor functions, how can you put transistors together to make a computer? In order to answer that question, you must start with a brief introduction to digital logic.

Begin with the examination of a single FET. In order to simplify the analysis, you can assume that zero voltage refers to a logic state of false, represented by a zero, and a negative voltage strong enough to turn the gate on refers to a logic state of 1, or true. The ouput of the device is given by current flow, where a zero represents no current flow (state=false), and a one represents current flow (state=true). The source is connected to zero bias and served as a reference.

P-N-P State Table		
G	**D**	**Current Flow**
0	0	0
1	0	0
0	1	0
1	1	1

As the table shows, the output of the FET, the current flow, is false unless both the gate and the drain are true. This type of analysis can be extended to more complex configurations of transistors.

13.13 Q: Build the truth table for the two-transistor structure shown below. Note that the ground symbol at the bottom of the schematic indicates zero voltage. A corresponding switch representation is shown to the right of the transistor schematic.

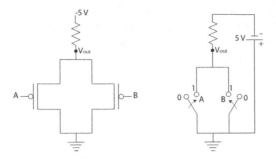

13.13 A: Building the truth table, you find that V_{out} is only true when both A and B are false.

A	B	Vout
0	0	1
0	1	0
1	0	0
1	1	0

This is known as a NOR (NOT OR) gate, and can be represented with the simpler digital logic schematic shown.

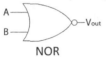

NOR

13.14 Q: Build the truth table for the two-transistor structure shown below.

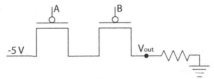

13.14 A: Building the truth table, you find that V_{out} is only true when both A and B are true.

A	B	Vout
0	0	0
0	1	0
1	0	0
1	1	1

This is known as an AND gate, and can be represented with the simpler digital logic schematic shown.

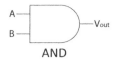

AND

13.15 Q: Build the truth table for the following configuration of a NOR gate where the two inputs are tied together.

13.15 A: Building the truth table, you find that V_{out} gives the opposite state of V_{in}.

V_{in}	Vout
0	1
1	0

By tying the inputs of a NOR gate together, you have created what is known as an inverter... whatever state goes in, the opposite state goes out. An inverter can be represented on a schematic diagram with the symbol shown below.

As you are beginning to see, using just these simple switches you can build more and more complex devices. But what about something more practical? How do you go from inverter, NOR, and AND gates to a computer? You continue to build more and more complex devices from these fundamental parts in incremental fashion. To provide a practical example, a half-adder is a device that performs very basic binary addition. It takes two binary inputs A and B (with values of either 0 or 1), and outputs the Sum of those of values, as shown in the following truth table. Note that if both inputs are a 1, the output, a binary two (1 0), includes not just a sum output, but also a Carry output to be "carried over" to the next mathematical operation.

Half Adder Truth Table			
A	**B**	**Sum**	**Carry**
0	0	0	0
0	1	1	0
1	0	1	0
1	1	0	1

Creating the logic for the "Sum" output can be done in a variety of ways us-ing a variety of gates. One combination using both inverter, AND, and NOR gates is shown below. See if you can verify the truth table above by walking through the logic functions using the device below.

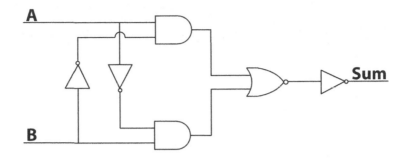

The logic for the Carry output is considerably simpler. Careful observation of the required output state given the input states indicates that the Carry can be made with a single AND gate.

Digital clocks, calculators, microwaves, radios, televisions, and computers are all run by integrated circuits consisting of hundreds, thousands, millions, and even billions of transistors connected in various configurations to execute a wide range of increasingly complex logic functions.

How, then, do you put millions and billions of these transistors onto a tiny piece of silicon the size of a fingernail? To answer that question, you must enter the realm of microelectronic fabrication!

Microelectronic Fabrication

The key to building billions of transistors on a tiny silicon wafer is to make them simultaneously in a process known as planar processing. Most micro-electronic facilities begin with thin, flat, single-crystal silicon wafers of di-ameters ranging from 100 to 450 millimeters, with thicknesses ranging from approximately 0.5 to 1 millimeter. These silicon wafers are near-perfect crystals, with purity better than 99.9999 percent.

The microelectronic devices are created through a series of steps which in-clude deposition of thin films of material, patterning of these thin films, se-lective etching of thin films, and modification of these materials. When these basic steps are repeated over and over, in a specific sequence, with varying parameters and patterns, hundreds of chips can be created on a single wafer, each chip containing hundreds of millions of functioning devices.

Looking at a single transistor in cross-section on the wafer, you can see it is made up of a p-n-p structure in the silicon, with a thin insulator above the n-type silicon region, known as the **gate dielectric**. Conducting contacts (typically heavily-doped poly-crystalline silicon or metals) form the source, gate, and drain regions.

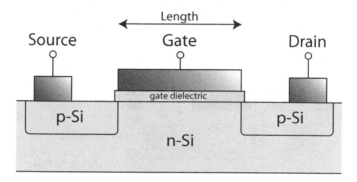

When a negative potential is placed on the conductor above the gate dielec-tric, the n-region in the silicon inverts to p-type and current can flow through the transistor. With no potential above the gate dielectric, the p-n-p struc-ture blocks current flow.

In modern devices, the gate region itself is often times less than 100 nanome-ters in length. These devices are embedded in the silicon and then connected together in a variety of configurations with alternating layers of insulators and conductors to form a wiring scheme on top of the transistors themselves.

In a modern integrated circuit processing plant, a state-of-the-art device may require more than 400 processing steps. The fabrication process uti-lizes the same equipment for different steps. Even so, due to the high tech nature of the equipment and extremely clean environment in which devices are manufactured, the factory may cost upwards of $5 billion to build, with approximately $1 billion for the building, and $4 billion in equipment!

Processing steps and equipment are broken down by function. Key process groups in a modern-day wafer fabrication plant, also known as a fab, include:

- Lithography - Create Patterns
- Etch and Clean - Transfer Patterns and Remove Contaminants
- Thin Film Deposition - Deposit Materials
- Implant and Diffusion - Modify Materials

Lithography

Lithography uses imaging techniques similar to photography to create tiny images on the silicon wafer. A thin layer sensitive to ultraviolet radiation known as **photoresist** is put on the wafer in a process known as spin coating. The wafer is placed in a bowl and a liquid solution of the photoresist is dispensed in a stream onto the wafer while the wafer spins at high speed, creating a very thin, uniform coating across the wafer.

The coated wafer is then placed in an illumination system known as a **stepper** or **scanner**. The stepper or scanner shines UV radiation through a lens, through a mask (part of circuit pattern), and into the layer of photoresist coating the wafer. When the UV rays strike the photoresist, they cause a chemical reaction to take place. This is called exposure, and exposed resist will become soluble or insoluble in certain chemicals.

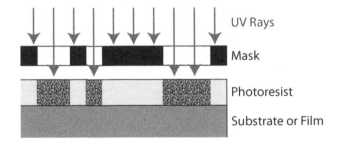

The wafer is then placed in a chemical developer, removing the regions of photoresist with high solubility and leaving the regions of photoresist with low solubility. In the example shown in the diagram, the exposed resist becomes less soluble. This is known as a negative photoresist process. Alternately, other types of photoresist become more soluble after exposure. Processes using this type of photoresist are known as positive photoresist processes.

So what's the big deal? Since the 1960s this process has been revised until you can transfer millions of patterns at once across a wafer, at feature sizes smaller than the wavelength of light.

Of course, this is simplified tremendously, and there are many variations on photoresist, developers, and steppers. Lithography is a rapidly evolving application of a variety of scientific disciplines, including, but not limited to, chemistry, optics, material science, computer programming, nuclear physics, and robotics.

ASML's state-of-the-art Twinscan NXT:1950i lithography system, capable of imaging 38 nanometer features while printing 200 wafers per hour. Image courtesy of ASML.

Etch and Clean

Etch and Clean (oftentimes subdivided into smaller process groups) transfer lithographic patterns into the wafer and remove contaminants and other materials from the wafer. Basic processes fall into two categories: those using wet (liquid) chemicals, and those using dry (gaseous) chemicals.

Wet clean processes involve the application of liquid chemicals to clean and remove a variety of unwanted materials from the wafer. Small particles of dust and debris from the atmosphere which land on the wafer can lead to non-functioning devices, and contamination sources may be passed from machine to machine. To provide an example, a dust particle in the air may have a diameter of 2 to 200 micrometers. The length of a transistor gate can be significantly less than 0.1 microns. Since a single speck of dust can cause a device not to function, devices are fabricated in extremely clean facilities known as cleanrooms.

Wafers will be subjected to clean processes many times during the fabrication cycle, each time with the goal of leaving a pristine surface and increasing the number of functional devices on each wafer (also known as the wafer yield).

In all cases, care must be taken to remove only the unwanted material, while leaving behind the desired materials and patterns on the wafer. This requires a solid background in chemistry, material science, and robotics.

Dry and wet etch processes utilize chemicals to selectively remove material, transferring a photoresist pattern into an underlying thin film. The process begins with a thin film that has been coated with photoresist and patterned by the lithography process.

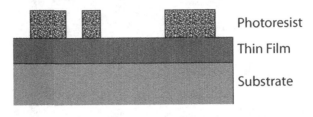

In a wet etch process, a liquid chemical which dissolves the thin film, but not the photoresist or substrate, is applied to transfer the pattern into the thin film.

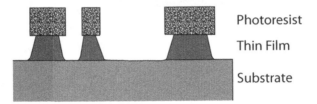

The photoresist pattern is then removed, leaving behind the patterned thin film. Note that the wet etch process is **isotropic**, meaning it removes material in both the horizontal and vertical direction. This results in angled sidewalls and wasted space where more transistors could be fabricated.

In a dry etch process, a gaseous chemical is placed into a very strong electric field, pulling the electrons from many of the gas molecules and creating gas ions in a type of **plasma** known as a **glow discharge**. By placing a negative potential underneath the wafer, the ions can be accelerated vertically downward toward the substrate. As the ions accelerate toward the substrate, they break up the gas molecules they move through into highly reactive species, while also providing both a physical attack and a chemical attack to the thin film to be removed. This type of dry etch process is therefore known as **reactive ion etching**, or RIE.

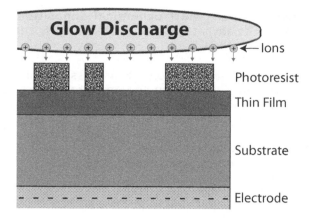

Through this effect, a dry etch process can create an **anisotropic** etch profile, meaning it removes materials in the vertical direction only.

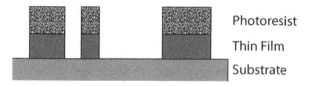

Once the thin film has been patterned, the photoresist is again removed.

The dry etch process results in less wasted space on the wafer, allowing for a higher packing density of transistors, and ultimately, a cheaper and faster IC.

Both wet and dry processes and equipment require great care and precision in the design and control of IC fabrication steps. This functional group is a great area for those with an interest in chemistry, material science, physics, engineering, mechanics, and robotics.

Thin Film Deposition

Thin Film Deposition processes place thin, uniform coatings of various materials onto wafers. This can be accomplished using a variety of methods, ranging from **physical vapor deposition** (PVD) processes such as **evaporation** and **sputtering** to chemical deposition processes such as **chemical**

vapor deposition (CVD) and **electrochemical deposition** (ECD). Each of these methods has its advantages and disadvantages, and is capable of depositing a variety of materials.

Evaporation is a physical vapor deposition process in which the wafer is placed in a chamber and subjected to very low pressures as the gases in the chamber are removed using a vacuum pump. Once the chamber is free of residual gases, the material to be deposited is subjected to a temperature sufficient to cause it to evaporate. The evaporated molecules are dispersed throughout the chamber, landing on the wafer (and the chamber walls) and condensing, coating the wafer uniformly. Evaporation is typically used for materials with high vapor pressures, such as metals, although it is also used occasionally for dielectric films such as SiO_2 and MgF_2, commonly used in the optics industry.

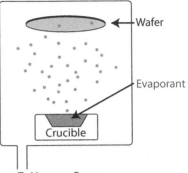

Sputtering is another physical vapor deposition process which occurs in a vacuum chamber. A large piece of the material to be deposited, known as a target, is bombarded with high energy argon ions from a glow discharge. When the argon ions strike the target, they knock off target atoms and molecules, which are then conveyed through the vacuum to the wafer, where they condense and form a thin film. Sputtering is most commonly used for depositing metal films, but, like evaporation, can also deposit insulating films with some slight process and equipment variations.

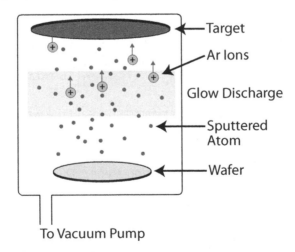

Chemical vapor deposition (CVD) can be accomplished in a vacuum (low pressure CVD, or LPCVD) or at atmospheric pressure (APCVD), although the highest quality films are deposited in a vacuum. Reactant gases are placed in a vacuum chamber with the wafer, and an energy source such as a glow discharge or infrared heating is used to activate a chemical reaction, causing the reactant gas molecules to break up, absorb on the wafer, and form a high quality thin film. CVD processes are typically used to deposit insulating and semiconducting films such as a silicon nitride (Si_3N_4), silicon dioxide (SiO_2), tungsten (W), and polycrystalline silicon (poly-Si), though it is also capable of depositing metallic films in a process known as metal-organic CVD (MOCVD). A special type of CVD, known as **atomic layer deposition**, or ALD, can even deposit a single monolayer of a film at a time in a tightly controlled process.

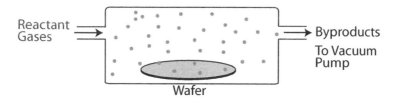

Electrochemical deposition (ECD) is used to deposit a thin conducting copper layer which will be patterned to form interconnects (integrated wires) connecting the various devices on an integrated circuit. The wafer is submerged upside down in a conducting copper sulfate solution. In the solution, a copper electrode is connected to a power supply, which is in turn connected to the wafer, forming a circuit through the conductive solution. The negative side of the power supply is connected to the wafer, forming the **cathode**. The positive side of the power supply is connected to the copper electrode in the solution, known as the **anode**.

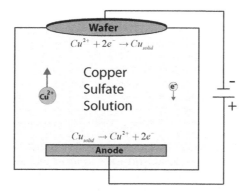

As current flows through the circuit (and through the solution), copper ions in solution near the cathode recombine with electrons from the current flow to form solid copper on the wafer.

$$Cu^{2+} + 2e^- \rightarrow Cu_{solid}$$

These copper ions are replaced by copper ions produced at the anode. The process is very carefully controlled, as the amount of copper deposited is directly proportional to the amount of current flowing through the circuit!

Implant and Diffusion

The **ion implantation** group is responsible for introducing donor and acceptor atoms into the silicon. The exact type of dopant atom, the depth to which it is placed, and the concentration at which it is introduced are all key facets which need to be well controlled and understood to successfully manufacture an IC.

An ion implanter is a device which takes a source material and creates an ion of the desired species. The species to be implanted is selected and accelerated through a column until it collides with the wafer, embedding itself in the silicon and replacing a silicon atom in the crystal lattice with the dopant atom.

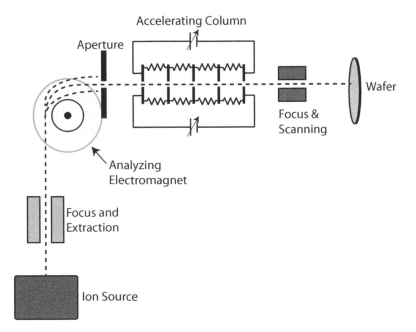

An ion source takes a gaseous species and creates a glow discharge, resulting in a large number of ions. Focus and extraction electrodes electrically pull the ions from the source into a beam. The moving ions then enter the mass analyzer, a large electromagnet which can be tuned to provide just the right force such that ions of the desired charge and mass pass through an output aperture, while ions of unwanted charge and mass are either bent too much or too little and are stopped by the **aperture**.

The remaining ions are then accelerated by a series of electrodes held at high potential, providing the ions with a tremendous amount of kinetic energy. The ions are then focused once more and scanned back and forth across a wafer, where they implant themselves in the silicon. If a photoresist mask is used prior to implantation, ions can be implanted in specific regions of the silicon.

Diffusion is the process by which particles spread from areas of higher concentration to lower concentration. In semiconductor manufacturing, diffusion processes are used to distribute ions which were introduced into the silicon by ion implantation throughout the silicon wafer. Because most p-type and n-type dopant ions diffuse quite slowly through silicon, the wafers are typically heated to temperatures ranging from 800 to 1200°C to speed up dopant distribution. This is accomplished through the use of a large, carefully controlled furnace designed specifically for wafer processing.

Diffusion can also modify the surface of the silicon. If silicon is placed in a high temperature furnace while in the presence of oxygen, the silicon will react with the oxygen to form a very high quality insulating layer of silicon dioxide (SiO_2), more commonly known as glass. Because these layers have extremely few imperfections, this can lead to even higher quality films than those deposited using PVD or CVD processes.

Applying these basic process steps to modify, pattern, deposit and etch various materials again and again and again, each time with different parameters and patterns, allows engineers to create billions of devices simultaneously. A state-of-the-art fabrication process may contain well in excess of 400 processing steps and 30 different masks.

Once the devices have been completed on the wafer, the wafers are tested, sent to the packaging group for dicing and encapsulation, tested again, and shipped.

13.16 Q: The thin insulator between the gate electrode and the silicon is known as the

(1) source region

(2) drain region

(3) gate dielectric

(4) p-channel

13.16 A: (3) gate dielectric

13.17 Q: In the lithographic process, the thin organic layer sensitive to ultraviolet radiation is known as

(1) photoresist

(2) polysilicon

(3) donor layer

(4) aperture

13.17 A: (1) photoresist

13.18 Q: Vertical etch profiles can be created by an (isotropic / anisotropic) process.

13.18 A: anisotropic

13.19 Q: Which process is used to deposit a single layer of atoms at a time?
(1) MOCVD
(2) PVD
(3) RIE
(4) ALD

13.19 A: (4) Atomic Layer Deposition (ALD)

Defining the actual microelectronic fabrication process to create a device is known as **process integration**. Process integration involves combining all of the individual silicon processing steps with the appropriate masks and equipment settings to create a functioning device. A simplified process flow to create an integrated resistor provides an indication of the methods in which these steps may be combined, and can be viewed online at:

http://www.aplusphysics.com/courses/honors/microe/integration.html.

Chapter 14: Waves & Sound

*"It would be possible to describe everything scientifically,
but it would make no sense; it would be without meaning,
as if you described a Beethoven symphony
as a variation of wave pressure."*

— Albert Einstein

Objectives

1. Define a pulse.
2. Describe the behavior of a pulse at a boundary.
3. Understand how the principle of superposition is applied when two pulses meet.
4. Define three terms to describe periodic waves: speed, wavelength, and frequency.
5. Explain the characteristics of transverse and longitudinal waves.
6. Describe the formation of standing waves.
7. Apply the principle of superposition to the phenomenon of interference.
8. Understand how resonance occurs.
9. Understand the nature of sound waves.
10. Apply the Doppler effect qualitatively to problems involving moving sources or moving observers.

Waves transfer energy through matter or space, and are found everywhere: sound waves, light waves, microwaves, radio waves, water waves, earthquake waves, slinky waves, x-rays, and on and on. Developing an understanding of waves will allow you to understand how energy is transferred in the universe, and will eventually lead to a better understanding of matter and energy itself!

Wave Characteristics

A **pulse** is a single disturbance which carries energy through a medium or through space. Imagine you and your friend holding opposite ends of a slinky. If you quickly move your arm up and down, a single pulse will travel down the slinky toward your friend.

If, instead, you generate several pulses at regular time intervals, you now have a wave carrying energy down the slinky. A **wave**, therefore is a re-peated disturbance which carries energy. The mass of the slinky doesn't move from one end of the slinky to the other, but the energy it carries does.

When a pulse or wave reaches a hard boundary, it reflects off the boundary, and is inverted. If a pulse or wave reaches a soft, or flexible, boundary, it still reflects off the boundary, but does not invert.

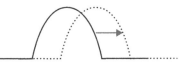

Waves can be classified in several different ways. One type of wave, known as a **mechanical wave**, requires a medium, or material, through which to travel. Ex-amples of mechanical waves include water waves, sound waves, slinky waves, and even seismic waves. **Electromagnetic waves**, on the other hand, do not require a medium in order to travel. Electromagnetic waves (or EM waves) are considered part of the Electro-magnetic Spectrum. Examples of EM waves include light, radio waves, microwaves, and even X-rays.

Further, waves can be classified based upon their direction of vibration. Waves in which the "particles" of the wave vibrate in the same direction as the wave velocity are known as **longitudinal**, or compressional, waves. Examples of longitudinal waves include sound waves and seismic P waves. Waves in which the particles of the wave vibrate perpendicular to the wave's direction of motion are known as **transverse** waves. Examples of transverse waves include seismic S waves, electromagnetic waves, and even stadium waves (the "human" waves you see at baseball and football games!).

Video animations of waves reflecting off boundaries as well as longitudinal and transverse waves can be viewed at http://bit.ly/gC1TMU.

Waves have a number of characteristics which define their behavior. Looking at a transverse wave, you can identify specific locations on the wave. The

---

highest points on the wave are known as **crests**. The lowest points on the wave are known as **troughs**. The **amplitude** of the wave, corresponding to the energy of the wave, is the distance from the baseline to a crest or the baseline to a trough.

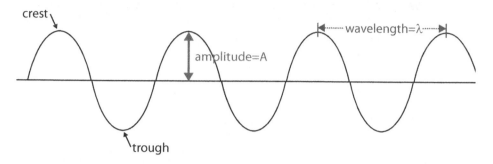

The length of the wave, or **wavelength**, represented by the Greek letter lambda (λ), is the distance between corresponding points on consecutive waves (i.e. crest to crest or trough to trough). Points on the same wave with the same displacement from equilibrium moving in the same direction (such as a crest to a crest or a trough to a trough) are said to be in phase (phase difference is 0° or 360°). Points with opposite displacements from equilibrium (such as a crest to a trough) are said to be 180° out of phase.

14.01 Q: Which type of wave requires a material medium through which to travel?

(1) sound

(2) television

(3) radio

(4) x ray

14.01 A: (1) sound is a mechanical wave and therefore requires a medium.

14.02 Q: The diagram below represents a transverse wave traveling to the right through a medium. Point A represents a particle of the medium.

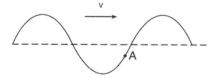

In which direction will particle A move in the next instant of time?

(1) up

(2) down

(3) left

(4) right

14.02 A: (2) particle A will move down as the wave passes.

14.03 Q: As a transverse wave travels through a medium, the individual particles of the medium move
(1) perpendicular to the direction of wave travel
(2) parallel to the direction of wave travel
(3) in circles
(4) in ellipses

14.03 A: (1) perpendicular to the direction of wave travel.

14.04 Q: A ringing bell is located in a chamber. When the air is removed from the chamber, why can the bell be seen vibrating but not be heard?
(1) Light waves can travel through a vacuum, but sound waves cannot.
(2) Sound waves have greater amplitude than light waves.
(3) Light waves travel slower than sound waves.
(4) Sound waves have higher frequency than light waves.

14.04 A: (1) Light is an EM wave, while sound is a mechanical wave.

14.05 Q: A single vibratory disturbance moving through a medium is called
(1) a node
(2) an antinode
(3) a standing wave
(4) a pulse

14.05 A: (4) a pulse.

14.06 Q: A periodic wave transfers
(1) energy, only
(2) mass, only
(3) both energy and mass
(4) neither energy nor mass

14.06 A: (1) energy, only.

14.07 Q: The diagram below represents a transverse wave.

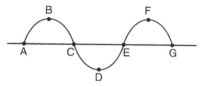

The wavelength of the wave is equal to the distance between points

(1) A and G
(2) B and F
(3) C and E
(4) D and F

14.07 A: (2) B and F is the wavelength as measured from crest to crest.

14.08 Q: The diagram below represents a periodic wave.

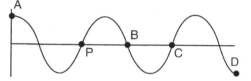

Which point on the wave is in phase with point P?

(1) A
(2) B
(3) C
(4) D

14.08 A: (3) Point C is the same point as point P but on a consecutive wave.

14.09 Q: The diagram below represents a transverse wave moving on a uniform rope with point A labeled. On the diagram, mark an X at the point on the wave that is 180° out of phase with point A.

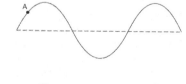

14.09 A:

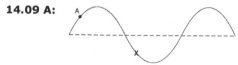

14.10 Q: The diagram below represents a transverse wave.

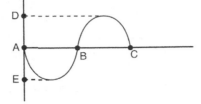

The distance between which two points identifies the amplitude of the wave?

(1) A and B

(2) A and C

(3) A and E

(4) D and E

14.10 A: (3) Amplitude is measured from the baseline to a crest or a trough, therefore amplitude is the distance between A and E.

14.11 Q: The diagram below represents a transverse wave traveling in a string.

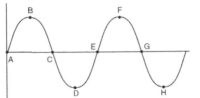

Which two labeled points are 180° out of phase?

(1) A and D

(2) B and F

(3) D and F

(4) D and H

14.11 A: (3) D and F.

In similar fashion, longitudinal waves also have amplitude and wavelength. In the case of longitudinal waves, however, instead of crests and troughs, the longitudinal waves have areas of high density (**compressions**) and areas of low density (**rarefactions**), as shown in the representation of the particles of a sound wave. The wavelength, then, of a compressional wave is the distance between compressions, or the distance between rarefactions. Once again, the amplitude corresponds to the energy of the wave.

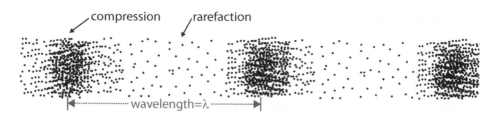

compression   rarefaction

wavelength=λ

14.12 Q: A periodic wave is produced by a vibrating
tuning fork. The amplitude of the wave
would be greater if the tuning fork were

(1) struck more softly

(2) struck harder

(3) replaced by a lower frequency tuning fork

(4) replaced by a higher frequency tuning fork

14.12 A: (2) Striking the tuning fork harder gives the tuning
fork more energy, increasing the sound wave's amplitude.

14.13 Q: Increasing the amplitude of a sound wave produces a sound with

(1) lower speed

(2) higher pitch

(3) shorter wavelength

(4) greater loudness

14.13 A: (4) greater loudness due to the greater energy / amplitude of the
wave.

14.14 Q: A longitudinal wave moves to the right through a uniform medium, as shown below.

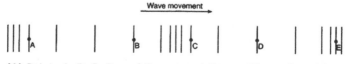

(A) Points A, B, C, D, and E represent the positions of particles of
the medium. What is the direction of the motion of the particles
at position C as the wave moves to the right?

(B) Between which two points on the wave could you measure a
complete wavelength?

14.14 A: (A) The particles move to the left and right at position C, as the
particles in a longitudinal wave vibrate parallel to the wave velocity.

(B) You could measure a complete wavelength between points
A and C, since A and C represent the same point on successive
waves.

The Wave Equation

The **frequency** (f) of a wave describes the number of waves that pass a given point in a time period of one second. The higher the frequency, the more waves that pass. Frequency is measured in number of waves per second (1/s), also known as a Hertz (Hz). If 60 waves pass a given point in a second, the frequency of the wave would be 60 Hz.

Closely related to frequency, the **period** (T) of a wave describes how long it takes for a single wave to pass a given point and can be found as the reciprocal of the frequency. Period is a measurement of time, and therefore is measured in seconds.

14.15 Q: What is the period of a 60-hertz electromagnetic wave traveling at 3.0×10^8 meters per second?

14.15 A: $T = \dfrac{1}{f} = \dfrac{1}{60\,Hz} = 0.0167s$

14.16 Q: Which unit is equivalent to meters per second?
(1) Hz•s
(2) Hz•m
(3) s/Hz
(4) m/Hz

14.16 A: (2) $\dfrac{m}{s} = Hz \bullet m$

14.17 Q: The product of a wave's frequency and its period is
(1) one
(2) its velocity
(3) its wavelength
(4) Planck's constant

14.17 A: (1) $f \bullet T = f \bullet \dfrac{1}{f} = 1$

Because waves move through space, they must have velocity. The velocity of a wave is a function of the type of wave and the medium it travels through. Electromagnetic waves moving through a vacuum, for instance, travel at roughly 3×10^8 m/s. This value is so famous and important in physics it is given its own symbol, **c**. When an electromagnetic wave enters a different medium, such as glass, it slows down. If the same wave were to then re-emerge from glass back into a vacuum, it would again travel at c, or 3×10^8 m/s.

The speed of a wave can be easily related to its frequency and wavelength. Speed of a wave is determined by the wave type and the medium it is traveling through. For a given wave speed, as frequency increases, wavelength must decrease, and vice versa. This can be shown mathematically using the wave equation.

$$v = f\lambda$$

14.18 Q: A periodic wave having a frequency of 5 hertz and a speed of 10 meters per second has a wavelength of

(1) 0.50 m

(2) 2.0 m

(3) 5.0 m

(4) 50. m

14.18 A: (2) $v = f\lambda$

$$\lambda = \frac{v}{f} = \frac{10 \, \text{m}/\text{s}}{5 Hz} = 2m$$

14.19 Q: If the amplitude of a wave is increased, the frequency of the wave will

(1) decrease

(2) increase

(3) remain the same

14.19 A: (3) remain the same.

14.20 Q: An electromagnetic wave traveling through a vacuum has a wavelength of 1.5×10^{-1} meters. What is the period of this electromagnetic wave?

(1) 5.0×10^{-10} s

(2) 1.5×10^{-1} s

(3) 4.5×10^7 s

(4) 2.0×10^9 s

14.20 A: (1) $v = f\lambda = \dfrac{\lambda}{T}$

$$T = \frac{\lambda}{v} = \frac{1.5 \times 10^{-1} m}{3 \times 10^{8} \, ^{m}\!/_{s}} = 5 \times 10^{-10} s$$

14.21 Q: A surfacing blue whale produces water wave crests having an amplitude of 1.2 meters every 0.40 seconds. If the water wave travels at 4.5 meters per second, the wavelength of the wave is

(1) 1.8 m

(2) 2.4 m

(3) 3.0 m

(4) 11 m

14.21 A: (1) $v = f\lambda$

$$\lambda = \frac{v}{f} = vT = (4.5 \, ^{m}\!/_{s})(0.4s) = 1.8m$$

Sound Waves

Sound is a mechanical wave which is observed by detecting vibrations in the inner ear. Typically, you think of sound as traveling through air, therefore the particles vibrating are air molecules. Sound can travel through other media as well, including water, wood, and even steel.

The particles of a sound wave vibrate in a direction parallel with the direction of the sound wave's velocity, therefore sound is a longitudinal wave. The speed of sound in air at standard temperature and pressure (STP) is 331 m/s.

14.22 Q: At an outdoor physics demonstration, a delay of 0.50 seconds was observed between the time sound waves left a loudspeaker and the time these sound waves reached a student through the air. If the air is at STP, how far was the student from the speaker?

14.22 A: $\overline{v} = \dfrac{d}{t} \rightarrow d = \overline{v}t = (331 \, ^{m}\!/_{s})(0.50s) = 166m$

14.23 Q: The sound wave produced by a trumpet has a frequency of 440 hertz. What is the distance between successive compressions in this sound wave as it travels through air at STP?

(1) 1.5×10^{-6} m

(2) 0.75 m

(3) 1.3 m

(4) 6.8×10^5 m

14.23 A: (2) $v = f\lambda$

$$\lambda = \frac{v}{f} = \frac{331\,^m/_s}{440\,Hz} = 0.75m$$

14.24 Q: A stationary research ship uses sonar to send a 1.18×10^3-hertz sound wave down through the ocean water. The reflected sound wave from the flat ocean bottom 324 meters below the ship is detected 0.425s second after it was sent from the ship.

(A) Calculate the speed of the sound wave in the ocean water.

(B) Calculate the wavelength of the sound wave in the water.

(C) Determine the period of the sound wave in the water.

14.24 A: (A) $\overline{v} = \frac{d}{t} = \frac{648m}{0.425s} = 1520\,^m/_s$

(B) $v = f\lambda$

$$\lambda = \frac{v}{f} = \frac{1525\,^m/_s}{1180\,Hz} = 1.29m$$

(C) $T = \frac{1}{f} = \frac{1}{1180\,Hz} = 8.47 \times 10^{-4}s$

When sound waves are observed through hearing, you pick up the amplitude, or energy, of the waves as loudness. The frequency of the wave is perceived as pitch, with higher frequencies observed as a higher pitch. Typically, humans can hear a frequency range of 20Hz to 20,000 Hz, although young observers can often detect frequencies above 20,000 Hz, an ability which declines with age.

Certain devices create strong sound waves at a single specific frequency. If another object, having the same "**natural frequency**," is impacted by these sound waves, it may begin to vibrate at this frequency, producing more sound waves. The phenomenon where one object emitting a sound wave with a specific frequency causes another object with the same natural frequency to vibrate is known as **resonance**. A dramatic demonstration of resonance involves an opera singer breaking a glass by singing a high pitch note. The singer creates a sound wave with a frequency equal to the natural frequency of the glass, causing the glass to vibrate at its natural, or resonant, frequency so energetically that it shatters. The same effect can be observed when you push someone on a swing. By pushing at the resonant frequency, the swing goes higher and higher!

14.25 Q: Sound waves strike a glass and cause it to shatter. This phenomenon illustrates

(1) resonance

(2) refraction

(3) reflection

(4) diffraction

14.25 A: (1) resonance

14.26 Q: A dampened fingertip rubbed around the rim of a crystal glass causes the glass to vibrate and produce a musical note. This effect is due to

(1) resonance

(2) refraction

(3) reflection

(4) rarefaction

14.26 A: (1) resonance

14.27 Q: Resonance occurs when one vibrating object transfers energy to a second object causing it to vibrate. The energy transfer is most efficient when, compared to the first object, the second object has the same natural

(1) frequency

(2) loudness

(3) amplitude

(4) speed

14.27 A: (1) frequency.

14.28 Q: A car traveling at 70 kilometers per hour accelerates to pass another car. When the car reaches a speed of 90 kilometers per hour the driver hears the glove compartment door start to vibrate. By the time the speed of the car is 100 kilometers per hour, the glove compartment door has stopped vibrating.
This vibrating phenomenon is an example of

(1) destructive interference

(2) the Doppler effect

(3) diffraction

(4) resonance

14.28 A: (4) resonance.

14.29 Q: Which wave phenomenon occurs when vibrations in one object cause vibrations in a second object?
(1) reflection
(2) resonance
(3) intensity
(4) tuning

14.29 A: (2) resonance.

Interference

When more than one wave travels through the same location in the same medium at the same time, the total displacement of the medium is governed by the principle of **superposition**. The principle of superposition simply states that the total displacement is the sum of all the individual displacements of the waves. The combined effect of the interaction of the multiple waves is known as **wave interference**.

14.30 Q: The diagram below shows two pulses approaching each other in a uniform medium. Diagram the superposition of the two pulses.

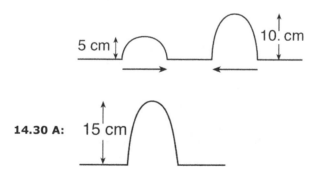

14.30 A:

When two or more pulses with displacements in the same direction interact, the effect is known as **constructive interference**. The resulting displacement is greater than the original individual pulses. Once the pulses have passed by each other, they continue along their original path in their original shape, as if they had never met.

When two or more pulses with displacements in opposite directions interact, the effect is known as **destructive interference**. The resulting displacements negate each other. Once the pulses have passed by each other, they continue along their original path in their original shape, as if they had never met. An animation of two pulses interfering constructively and destructively is available at http://bit.ly/hyJ3lZ.

14.31 Q: The diagram below represents two pulses approaching each other from opposite directions in the same medium.

Which diagram best represents the medium after the pulses have passed through each other?

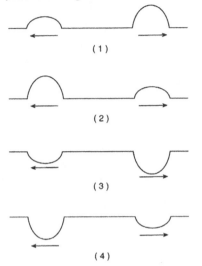

14.31 A: (2) the pulses continue as if they had never met.

14.32 Q: The diagram below represents shallow water waves of constant wavelength passing through two small openings, A and B, in a barrier.

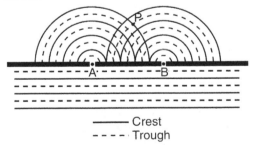

———— Crest
- - - - Trough

Which statement best describes the interference at point P?

(1) It is constructive, and causes a longer wavelength.

(2) It is constructive, and causes an increase in amplitude.

(3) It is destructive, and causes a shorter wavelength.

(4) It is destructive, and causes a decrease in amplitude.

14.32 A: (4) when a crest and a trough meet, destructive interference causes a decrease in amplitude.

14.33 Q: The diagram below shows two pulses of equal amplitude, A, approaching point P along a uniform string.

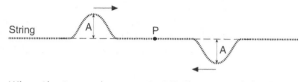

When the two pulses meet at P, the vertical displacement of the string at P will be

(1) A

(2) 2A

(3) 0

(4) A/2

14.33 A: (3) the pulses will experience destructive interference.

14.34 Q: The diagram below represents two pulses approaching each other.

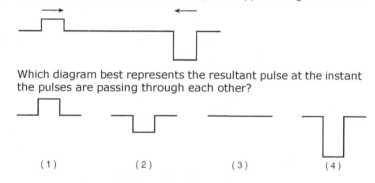

Which diagram best represents the resultant pulse at the instant the pulses are passing through each other?

(1)   (2)   (3)   (4)

14.34 A: (2) shows the superposition (addition) of the two pulses.

14.35 Q: Two waves having the same amplitude and frequency are traveling in the same medium. Maximum destructive interference will occur when the phase difference between the waves is

(1) 0°

(2) 90°

(3) 180°

(4) 270°

14.35 A: (3) Maximum destructive interference occurs at a phase difference of 180°.

Standing Waves

When waves of the same frequency and amplitude traveling in opposite directions meet, a standing wave is produced. A **standing wave** is a wave in which certain points (**nodes**) appear to be standing still and other points (**antinodes**) vibrate with maximum amplitude above and below the axis.

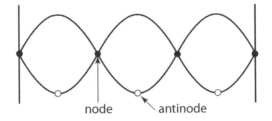

node   antinode

Looking at the standing wave produced above, you can see a total of four nodes in the wave, and three antinodes. For any standing wave pattern, you will always have one more node than antinode.

Standing waves can be observed in a variety of patterns and configurations, and are responsible for the functioning of most musical instruments. Guitar strings, for example, demonstrate a standing wave pattern. By fretting the strings, you adjust the wavelength of the string, and therefore the frequency of the standing wave pattern, creating a different pitch. Similar functionality is seen in instruments ranging from pianos and drums to flutes, harps, trombones, xylophones, and even pipe organs!

14.36 Q: While playing, two children create a standing wave in a rope, as shown in the diagram below.

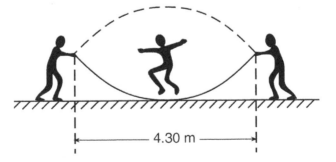

4.30 m

A third child participates by jumping the rope. What is the wavelength of this standing wave?

(1) 2.15 m

(2) 4.30 m

(3) 6.45 m

(4) 8.60 m

14.36 A: (4) the standing wave shown is half a wavelength, therefore the total wavelength must be 8.6m.

14.37 Q: Wave X travels eastward with frequency f and amplitude A. Wave Y, traveling in the same medium, interacts with wave X and produces a standing wave. Which statement about wave Y is correct?

(1) Wave Y must have a frequency of f, an amplitude of A, and be traveling eastward.

(2) Wave Y must have a frequency of 2f, an amplitude of 3A, and be traveling eastward.

(3) Wave Y must have a frequency of 3f, an amplitude of 2A, and be traveling westward.

(4) Wave Y must have a frequency of f, an amplitude of A, and be traveling westward.

14.37 A: (4) Standing waves are created when waves with the same frequency and amplitude traveling in opposite directions meet.

14.38 Q: The diagram below represents a wave moving toward the right side of this page.

Which wave shown below could produce a standing wave with the original wave?

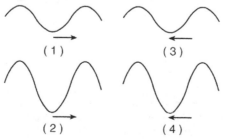

14.38 A: (3) must have same frequency, amplitude, and be traveling in the opposite direction in the same medium.

14.39 Q: The diagram below shows a standing wave.

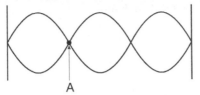

A

Point A on the standing wave is

(1) a node resulting from constructive interference

(2) a node resulting from destructive interference

(3) an antinode resulting from constructive interference

(4) an antinode resulting from destructive interference

14.39 A: (2) a node resulting from destructive interference.

14.40 Q: One end of a rope is attached to a variable speed drill and the other end is attached to a 5-kilogram mass. The rope is draped over a hook on a wall opposite the drill. When the drill rotates at a frequency of 20 Hz, standing waves of the same frequency are set up in the rope. The diagram below shows such a wave pattern.

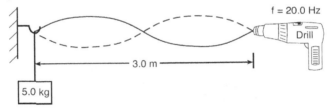

(A) Determine the wavelength of the waves producing the standing wave pattern.

(B) Calculate the speed of the wave in the rope.

14.40 A: (A) Wavelength is 3.0 meters from diagram.

(B) $v = f\lambda = (20\,Hz)(3m) = 60\,{}^m\!/_s$

Due to their very nature, waves exhibit a number of behaviors that may not be obvious upon first inspection, including the Doppler Effect, reflection, refraction, and diffraction. Understanding these behaviors brings mankind closer to understanding the universe, while also providing a number of useful applications including, but not limited to, radar, sonography, digital televisions, mirrors, telescopes, glasses, contact lenses, atomic research, and even holography!

Doppler Effect

The shift in a wave's observed frequency due to relative motion between the source of the wave and the observer is known as the **Doppler Effect**. In essence, when the source and/or observer are moving toward each other, the observer perceives a shift to a higher frequency, and when the source and/or observer are moving away from each other, the observer perceives a lower frequency.

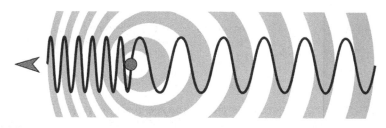

This can be observed when a vehicle travels past you. As you hear the vehicle approach, you can observe a higher frequency noise, and as the vehicle passes by you and then moves away, you observe a lower frequency noise. This effect is the principle behind radar guns to measure an object's speed as well as meteorology radar which provides data on wind speeds.

The Doppler Effect results from waves having a fixed speed in a given medium. As waves are emitted, a moving source or observer encounters the wave fronts at a different frequency than the waves are emitted, resulting in a perceived shift in frequency. The video and animation at http://bit.ly/epLkPj may help you visualize this effect.

14.41 Q: A car's horn produces a sound wave of constant frequency. As the car speeds up going away from a stationary spectator, the sound wave detected by the spectator

(1) decreases in amplitude and decreases in frequency

(2) decreases in amplitude and increases in frequency

(3) increases in amplitude and decreases in frequency

(4) increases in amplitude and increases in frequency

14.41 A: (1) decreases in amplitude because the distance between source and observe is increasing, and decreases in frequency because the source is moving away from the observer.

14.42 Q: A car's horn is producing a sound wave having a constant frequency of 350 hertz. If the car moves toward a stationary observer at constant speed, the frequency of the car's horn detected by this observer may be

(1) 320 Hz

(2) 330 Hz

(3) 350 Hz

(4) 380 Hz

14.42 A: (4) If source is moving toward the stationary observer, the observed frequency must be higher than source frequency.

14.43 Q: A radar gun can determine the speed of a moving automobile by measuring the difference in frequency between emitted and reflected radar waves. This process illustrates

(1) resonance

(2) the Doppler effect

(3) diffraction

(4) refraction

14.43 A: (2) the Doppler effect.

14.44 Q: The vertical lines in the diagram represent compressions in a sound wave of constant frequency propagating to the right from a speaker toward an observer at point A.

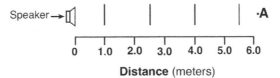

Speaker →

0   1.0   2.0   3.0   4.0   5.0   6.0

Distance (meters)

(A) Determine the wavelength of this sound wave.

(B) The speaker is then moved at constant speed toward the observer at A. Compare the wavelength of the sound wave received by the observer while the speaker is moving to the wavelength observed when the speaker was at rest.

14.44 A: (A) Wavelength is compression to compression, or 1.5m.

(B) Observed frequency is higher while speaker is moving toward the observer due to the Doppler Effect, so the observed wavelength must be shorter.

14.45 Q: A student sees a train that is moving away from her and sounding its whistle at a constant frequency. Compared to the sound produced by the whistle, the sound observed by the student is

(1) greater in amplitude

(2) a transverse wave rather than a longitudinal wave

(3) higher in pitch

(4) lower in pitch

14.45 A: (4) lower in pitch since the source is moving away from the observer.

An exciting application of the Doppler Effect involves the analysis of radiation from distant stars and galaxies in the universe. Based on the basic elements that compose stars, scientists know what frequencies of radiation to look for. However, when analyzing these objects, they observe frequencies shifted toward the red end of the electromagnetic spectrum (lower frequencies), known as the **Red Shift**. This indicates that these celestial objects must be moving away from the Earth. The more distant the object, the greater the red shift. Putting this together, you can conclude that more distant celestial objects are moving away from Earth faster than nearer objects, and therefore, the universe must be expanding!

14.46 Q: When observed from Earth, the wavelengths of light emitted by a star are shifted toward the red end of the electromagnetic spectrum. This redshift occurs because the star is

(1) at rest relative to Earth

(2) moving away from Earth

(3) moving toward Earth at decreasing speed

(4) moving toward Earth at increasing speed

14.46 A: (2) moving away from Earth.

Chapter 15: Optics

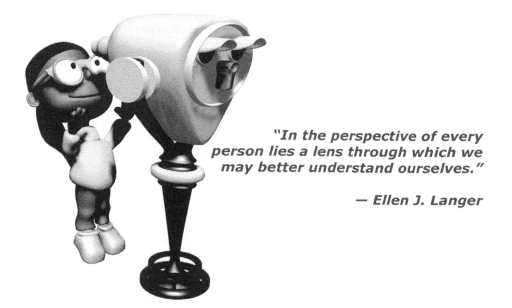

"In the perspective of every person lies a lens through which we may better understand ourselves."

— Ellen J. Langer

Objectives

1. Explain the law of reflection.
2. Analyze the behavior of plane and curved mirrors.
3. Understand and apply Snell's law.
4. Calculate the index of refraction in a medium.
5. Determine the critical angle for an interface.
6. Analyze the behavior of convex and concave lenses.
7. Relate the diffraction of light to its wave characteristics.
8. Describe Young's double-slit experiment.
9. Recognize characteristics of EM waves and determine the type of EM wave based on its characteristics.

Reflection

When a wave hits a boundary, three different events can occur. The wave may be:

- Reflected - wave bounces off a boundary
- Transmitted - wave is transmitted into the new medium
- Absorbed - energy of the wave is transferred into the boundary medium

The **law of reflection** states that the angle at which a wave strikes a reflective medium (the **angle of incidence**, or θ_i) is equal to the angle at which a wave reflects off the medium (the **angle of reflection**, or θ_r). Put more simply, $\theta_i = \theta_r$. In all cases, the angle of incidence and the angle of reflection are measured from a line perpendicular, or normal, to the reflecting surface.

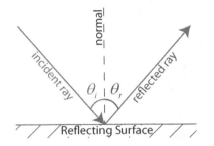

Although all waves can exhibit these behaviors, electromagnetic light waves are typically considered for demonstration purposes. When a wave bounces off a reflective surface, the nature of its reflection depends largely on the nature of the surface. Rough surfaces tend to reflect light in a variety of directions in a process known as diffuse reflection. **Diffuse reflection** is the type of reflection typically observed off of pieces of paper. Smooth surfaces tend to reflect light waves in a more regular fashion, such that the reflected rays maintain their parallelism. This process is known as **specular reflection**, and is commonly observed in mirrors.

15.1 Q: The diagram below represents a light ray striking the boundary between air and glass.

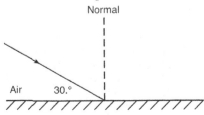

What would be the angle between this light ray and its reflected ray?

(1) 30°

(2) 60°

(3) 120°

(4) 150°

15.1 A: (3) recall that ray angles are always measured to the normal, so the angle between the two rays is 60°+60°=120°.

15.2 Q: A sonar wave is reflected from the ocean floor. For which angles of incidence do the wave's angle of reflection equal its angle of incidence?

(1) angles less than 45°, only

(2) an angle of 45°, only

(3) angles greater than 45°, only

(4) all angles of incidence

15.2 A: (4) the law of reflection applies to all types of waves reflecting off a surface.

Mirrors

When you look in a flat (plane) mirror, you see a reflection, or an image, of an object. Light rays from the object reach the plane mirror and are reflected back to the observer, creating an image of the object. The image is known as a **virtual image** because the reflected rays don't actually pass through the image. All virtual images are upright.

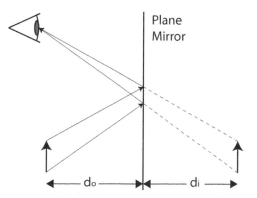

The distance from the object to the mirror is known as the **object distance** (d_o), and the distance from the image to the mirror is known as the **image distance** (d_i). For virtual images, the image distance is negative. The **magnification** of an image is found using the magnification equation, which relates the image and object distances to the image (h_i) and object (h_o) heights.

$$m = \frac{-d_i}{d_o} = \frac{h_i}{h_o}$$

In the case of a plane mirror, the magnitude of the image distance is equal to the magnitude of the object distance, therefore the image appears the same size as the object.

15.3 Q: A student stands 2 meters in front of a vertical plane mirror. As the student walks toward the mirror, the image

(1) decreases in magnification and remains virtual

(2) decreases in magnification and remains real

(3) remains the same magnification and remains virtual

(4) remains the same magnification and remains real

15.3 A: (3) remains the same magnification and remains virtual since the image stays behind the reflecting surface, and the magnification of a plane mirror is always 1.

15.4 Q: Which diagram best represents image *I*, which is formed by placing object *O* in front of a plane mirror?

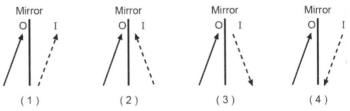

(1) (2) (3) (4)

15.4 A: (2)

15.5 Q: In the diagram below, a person is standing 5 meters from a plane mirror. The chair in front of the person is located 2 meters from the mirror.

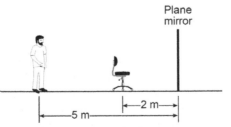

What is the distance between the person and the image he observes of the chair?

15.5 A: 7m

Not all mirrors are plane mirrors, however. The inner surface of a spherical **concave mirror** is reflective. Light rays coming into a mirror parallel to the **principal** axis (a virtual line perpendicular to the mirror's surface) are reflected from the plane of the mirror and converge through the **focal point** of the mirror. Concave mirrors are also known as converging mirrors. The focal point of a spherical mirror is half its radius of curvature.

Light rays passing through the center of curvature strike the mirror and are reflected back through the center of curvature. Light rays from the object passing directly through the focal point are reflected back parallel to the principal axis. The convergence of the reflected rays creates an image. The distance from the focal point to the mirror's surface is known as the focal distance (f). The image is known as a **real image** because the reflected rays pass through the image. Real images are inverted.

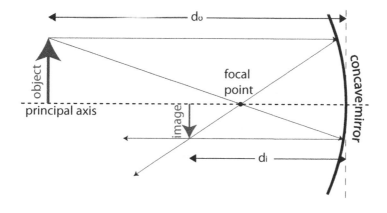

The relationship between the focal distance, the object distance, and the image distance is described by the mirror equation, also known as the lensmaker's equation. By convention, object and image distances are positive on the reflecting side of the mirror, and negative on the non-reflecting side of the mirror.

$$\frac{1}{f} = \frac{1}{d_o} + \frac{1}{d_i}$$

Analyzing an object inside the focal point of a concave mirror requires the same basic procedure. In this case, however, the reflected rays diverge on the reflective side of the mirror. To find the image, you must extend the real reflected rays back through the mirror onto the non-reflective side. The convergence of these extended rays leads to an upright, virtual image.

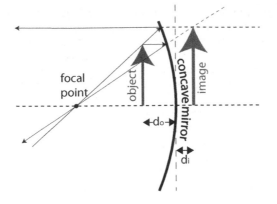

15.6 Q: An incident light ray travels parallel to the principal axis of a concave spherical mirror. After reflecting from the mirror, the light ray will travel

(1) through the mirror's focal point

(2) through the mirror's center of curvature

(3) parallel to the mirror's principal axis

(4) normal to the mirror's principal axis

15.6 A: (1) light rays parallel to the principal axis are reflected through the mirror's focal point.

15.7 Q: The diagram below shows an object located at point P, 0.25 meters from a concave spherical mirror with focal point F. The focal length of the mirror is 0.10 meters.

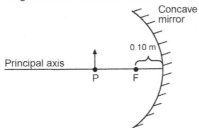

How does the image change as the object is moved from point P toward point F?

(1) Its distance from the mirror decreases and the size of the image decreases.

(2) Its distance from the mirror decreases and the size of the image increases.

(3) Its distance from the mirror increases and the size of the image decreases.

(4) Its distance from the mirror increases and the size of the image increases.

15.7 A: (4) Using the mirror equation at right, you observe that by moving the object toward point F, the object distance d_o decreases. Since the focal point f is fixed, the image distance d_i must increase, so the image's distance from the mirror increases. Further, using the magnification equation, you observe that d_i increasing and d_o decreasing leads to an increase in magnification m.

$$\frac{1}{f} = \frac{1}{d_o} + \frac{1}{d_i}$$

$$m = \frac{-d_i}{d_o} = \frac{h_i}{h_o}$$

15.8 Q: An object arrow is placed in front of a concave mirror having center of curvature C and focal point F. Which diagram best shows the location of point I, the image of the tip of the object arrow? Is the image real or virtual? Upright or inverted?

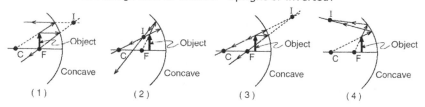

(1)   (2)   (3)   (4)

15.8 A: (3) The image is virtual and upright.

15.9 Q: A candle is placed 0.24 meters in front of a converging mirror that has a focal length of 0.12 meters. How far from the mirror is the image of the candle located?

(1) 0.08 m   (2) 0.12 m

(3) 0.24 m   (4) 0.36 m

15.9 A: (3) $\dfrac{1}{f} = \dfrac{1}{d_o} + \dfrac{1}{d_i} \rightarrow \dfrac{1}{d_i} = \dfrac{1}{f} - \dfrac{1}{d_o} \rightarrow$

$$\frac{1}{d_i} = \frac{1}{.12m} - \frac{1}{.24m} \rightarrow \frac{1}{d_i} = 4.17m^{-1} \rightarrow$$

$$d_i = \frac{1}{4.17m^{-1}} = 0.24m$$

The outer surface of a spherical **convex mirror** is reflective. Light rays coming into a convex mirror parallel to the principal axis are reflected away from the principal axis on a virtual line connecting the point of contact with the mirror plane and the focal point on the **non-reflecting** side of the mirror. For this reason, convex mirrors are also known as diverging mirrors. Light rays which strike the center of the mirror are reflected at the same angle. Because the light rays never converge on the reflective side of a convex mirror, convex mirrors only produce virtual images that are upright and reduced in size.

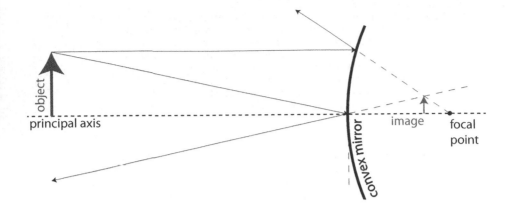

15.10 Q: Which optical device causes parallel light rays to diverge?
  (1) convex mirror   (2) plane mirror
  (3) concave mirror   (4) convex lens

15.10 A: (1) convex mirrors are also known as diverging mirrors.

15.11 Q: Images formed by diverging (convex) mirrors are always
  (1) real and inverted
  (2) real and erect
  (3) virtual and inverted
  (4) virtual and erect

15.11 A: (4) Images formed by diverging mirrors are virtual, erect, and reduced in size.

15.12 Q: Light rays from a candle flame are incident on a convex mirror. After reflecting from the mirror, these light rays
  (1) converge and form a virtual image
  (2) converge and form a real image
  (3) diverge and form a virtual image
  (4) diverge and form a real image

15.12 A: (3) Light rays incident on convex mirrors diverge and form only virtual images.

15.13 Q: The radius of curvature of a spherical mirror is R. The focal length of this mirror is equal to

(1) R/2　　　　　　　　　　(2) 2R

(3) R/4　　　　　　　　　　(4) 4R

15.13 A: (1) The focal length of a mirror is equal to half its radius of curvature.

Refraction

When a wave reaches the boundary between media, part of the wave is reflected and part of the wave enters the new medium. As the wave enters the new medium, the speed of the wave changes, and the frequency of a wave remains constant, therefore, consistent with the wave equation, $v=f\lambda$, the wavelength of the wave must change.

15.14 Q: When a wave enters a new material, what happens to its speed, frequency, and wavelength?

15.14 A: Speed changes, frequency remains constant, and wavelength changes.

The front of a wave has some actual width, and if the wave does not impinge upon the boundary between media at a right angle, not all of the wave enters the new medium and changes speed at the same time. This causes the wave to bend as it enters a new medium in a process known as **refraction**.

To better illustrate this, imagine you're in a line in a marching band, connected with your bandmates as you march at a constant speed down the field in imitation of a wave front. As your wavefront reaches a new medium that slows you down, such as a mud pit, the band members reaching the mud pit slow down before those who reach the pit later. Since you are all connected in a wave front, the entire wave shifts directions (refracts) as it passes through the boundary between field and mud!

The **index of refraction** (n) is a measure of how much light slows down in a material. In a vacuum and in air, all electromagnetic waves have a speed of $c=3 \times 10^8$ m/s. In other materials, light slows down. The ratio of the speed of light in a vacuum to the speed of light in the new material is known as the index of refraction (n). The slower the wave moves in the material, the larger the index of refraction:

$$n = \frac{c}{v}$$

Not only does index of refraction depend upon the medium the light wave is traveling through, it also varies with frequency. Thankfully, its variation is typically fairly small.

Absolute Indices of Refraction	
$(f = 5.09 \times 10^{14}$ Hz$)$	
Air	1.00
Corn oil	1.47
Diamond	2.42
Ethyl alcohol	1.36
Glass, crown	1.52
Glass, flint	1.66
Glycerol	1.47
Lucite	1.50
Quartz, fused	1.46
Sodium chloride	1.54
Water	1.33
Zircon	1.92

15.15 Q: A light ray traveling in air enters a second medium and its speed slows to 1.71 x 10⁸ m/s. What is the absolute index of refraction of the second medium?

15.15 A: $n = \dfrac{c}{v} = \dfrac{3 \times 10^{8} \; m/s}{1.71 \times 10^{8} \; m/s} = 1.75$

15.16 Q: In which way does blue light change as it travels from diamond into crown glass?
(1) Its frequency decreases.
(2) Its frequency increases.
(3) Its speed decreases.
(4) Its speed increases.

15.16 A: (4) Its speed increases because it crosses from a higher index of refraction material to a lower index of refraction material.

15.17 Q: Which characteristic is the same for every color of light in a vacuum?

(1) energy

(2) frequency

(3) speed

(4) period

15.17 A: (3) the speed of all EM waves in a vacuum is 3.0×10^8 m/s.

15.18 Q: A periodic wave travels at speed v through medium A. The wave passes with all its energy into medium B. The speed of the wave through medium B is v/2. Draw the wave as it travels through medium B.

15.18 A:

15.19 Q: A beam of monochromatic light has a wavelength of 5.89×10^{-7} meters in air. Calculate the wavelength of this light in diamond.

15.19 A:
$$\frac{n_2}{n_1} = \frac{\lambda_1}{\lambda_2}$$

$$\lambda_2 = \frac{n_1 \lambda_1}{n_2} = \frac{(1.00)(5.89 \times 10^{-7}\,m)}{2.42} = 2.43 \times 10^{-7}\,m$$

15.20 Q: The speed of light in a piece of plastic is 2.00×10^8 meters per second. What is the absolute index of refraction of this plastic?

(1) 1.00

(2) 0.670

(3) 1.33

(4) 1.50

15.20 A: (4) $n = \dfrac{c}{v} = \dfrac{3 \times 10^8\,m/s}{2 \times 10^8\,m/s} = 1.5$

The amount a light wave bends as it enters a new medium is given by the law of refraction, also known as **Snell's Law**, which states:

$$n_1 \sin \theta_1 = n_2 \sin \theta_2$$

In this formula, n_1 and n_2 are the indices of refraction of the two media, and θ_1 and θ_2 correspond to the angles of the incident and refracted rays, again measured from the normal. Light bends toward the normal as it enters a material with a higher index of refraction (slower material), and bends away from the normal as it enters a material with a lower index of refraction (faster material).

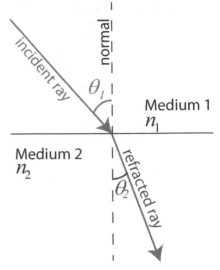

15.21 Q: A ray of light (f=5.09×10¹⁴ Hz) traveling in air is incident at an angle of 40° on an air-crown glass interface as shown below.

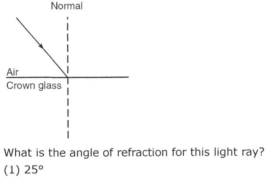

What is the angle of refraction for this light ray?
(1) 25°
(2) 37°
(3) 40°
(4) 78°

15.21 A: (1) $n_1 \sin \theta_1 = n_2 \sin \theta_2$

$$\theta_2 = \sin^{-1}\left(\frac{n_1 \sin \theta_1}{n_2}\right) = \sin^{-1}\left(\frac{1.00 \times \sin 40^\circ}{1.52}\right) = 25^\circ$$

15.22 Q: A ray of monochromatic light (f =5.09×10¹⁴ Hz) passes from air into Lucite at an angle of incidence of 30°.

(A) Calculate the angle of refraction in the Lucite.

(B) Using a protractor and straightedge, draw the refracted ray in the Lucite.

15.22 A: (A) $n_1 \sin\theta_1 = n_2 \sin\theta_2$

$$\theta_2 = \sin^{-1}\left(\frac{n_1 \sin\theta_1}{n_2}\right) = \sin^{-1}\left(\frac{1.00 \times \sin 30°}{1.50}\right) = 19°$$

(B)

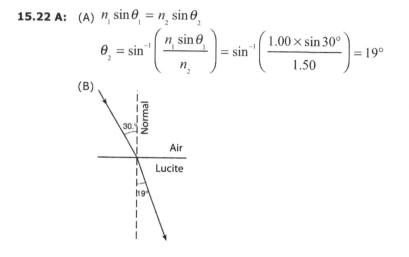

15.23 Q: When a light wave enters a new medium and is refracted, there must be a change in the light wave's

(1) color

(2) frequency

(3) period

(4) speed

15.23 A: (4) the change in a wave's speed causes its refraction.

15.24 Q: A ray of light (f=5.09×10¹⁴ Hz) traveling in air strikes a block of sodium chloride at an angle of incidence of 30°. What is the angle of refraction for the light ray in the sodium chloride?

(1) 19°

(2) 25°

(3) 40°

(4) 49°

15.24 A: (1) $n_1 \sin\theta_1 = n_2 \sin\theta_2$

$$\theta_2 = \sin^{-1}\left(\frac{n_1 \sin\theta_1}{n_2}\right) = \sin^{-1}\left(\frac{1.00 \times \sin 30°}{1.54}\right) = 19°$$

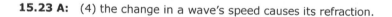

Which diagram best represents the behavior of a ray of mono-chromatic light in air incident on a block of crown glass?

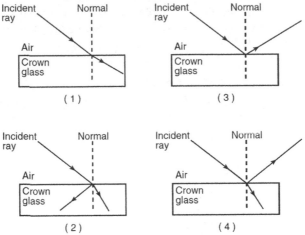

15.25 A: (4) shows both reflection and refraction of the incoming light ray.

Thin Lenses

A common application of refraction is the optical lens. Much like mirrors, lenses come in two types: convex and concave. When working with lenses, however, convex lenses are converging lenses, and concave lenses are diverging lenses.

In the case of lenses, similar rules for ray tracing apply. For a convex lens, a ray parallel to the principal axis is refracted through the far focal point of the lens. In addition, a ray drawn from the object through the center of the lens passes through the center of the lens unbent.

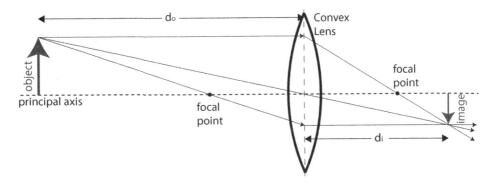

15.26 Q: The diagram below shows an arrow placed in front of a converging lens.

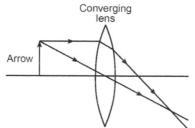

Converging lens

Arrow

The lens forms an image of the arrow that is

(1) real and inverted

(2) real and erect

(3) virtual and inverted

(4) virtual and erect

15.26 A: (1) the image will be real and inverted.

15.27 Q: An object is located 0.15 meters from a converging lens with focal length 0.10 meters. How far from the lens is the image formed?

15.27 A:

$$\frac{1}{f} = \frac{1}{d_o} + \frac{1}{d_i} \rightarrow \frac{1}{d_i} = \frac{1}{f} - \frac{1}{d_o} \rightarrow$$

$$\frac{1}{d_i} = \frac{1}{.10m} - \frac{1}{.15m} \rightarrow \frac{1}{d_i} = 3.33m^{-1} \rightarrow$$

$$d_i = \frac{1}{3.33m^{-1}} = 0.3m$$

15.28 Q: A converging lens forms a real image that is four times larger than the object. If the image is located 0.16 meters from the lens, what is the object distance?

15.28 A: First, realize that since the image is real, the image distance, by convention, is negative. Therefore, you are given the magnification of m=4, and the image distance d_i=-0.16m. Now you can apply the magnification equation to solve for the object distance d_o.

$$m = \frac{-d_i}{d_o} \rightarrow d_o = \frac{-d_i}{m} = \frac{-(-0.16m)}{4} = 0.04m$$

For a concave lens, a ray from the object parallel to the principal axis is refracted away from the principal axis on a line from the near focal point through the point where the ray intercepts the center of the lens. In addition, any ray that passes from the object through a focal point is refracted parallel to the principal axis. This leads to upright, virtual, reduced images from concave diverging lenses.

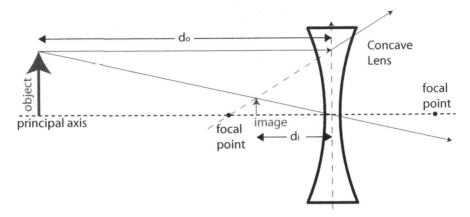

The same equations used for analysis of mirrors also apply to the analysis of thin lenses. Recall the lensmaker's equation:

$$\frac{1}{f} = \frac{1}{d_o} + \frac{1}{d_i}$$

Calculation of the magnification of a lens also uses the same equation used in the analysis of mirrors:

$$m = \frac{-d_i}{d_o} = \frac{h_i}{h_o}$$

15.29 Q: In the diagram below, parallel light rays in air diverge as a result of interacting with an optical device.

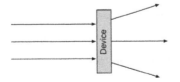

The device could be a

(1) convex glass lens

(2) rectangular glass block

(3) plane mirror

(4) concave glass lens

15.29 A: (4) a concave lens is also known as a diverging lens.

15.30 Q: Which glass lens in air can produce an enlarged real image of an object?

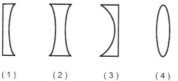

(1)   (2)   (3)   (4)

15.30 A: (4) convex lenses can produce enlarged real images.

15.31 Q: Which ray best represents the path of light ray R after it passes through the lens?

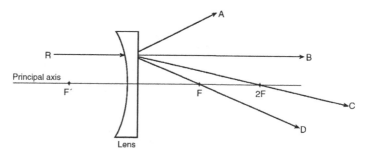

15.31 A: (A) shows the diverging light ray from a diverging (concave) lens.

15.32 Q: The diagram shows an object placed between 1 and 2 focal lengths from a converging lens. The image of the object produced by the lens is

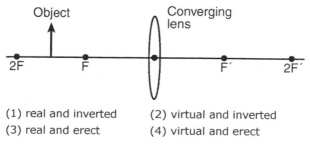

(1) real and inverted   (2) virtual and inverted

(3) real and erect   (4) virtual and erect

15.32 A: (1) real and inverted

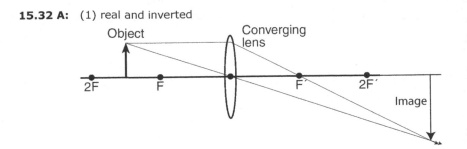

15.33 Q: A converging lens has a focal length of 0.80 meters. A light ray travels from the object to the lens parallel to the principal axis.

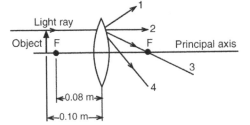

(A) Which line best represents the path of the ray after it leaves the lens?

(B) How far from the lens is the image formed?

(C) Which phenomenon best explains the path of the light ray through the lens?

(1) diffraction   (2) dispersion

(3) reflection   (4) refraction

15.33 A: (A) (3) Rays parallel to the principal axis are refracted through the far focal point of a convex lens.

(B) $\dfrac{1}{f} = \dfrac{1}{d_o} + \dfrac{1}{d_i} \rightarrow \dfrac{1}{d_i} = \dfrac{1}{f} - \dfrac{1}{d_o} \rightarrow$

$\dfrac{1}{d_i} = \dfrac{1}{0.08m} - \dfrac{1}{0.10m} \rightarrow \dfrac{1}{d_i} = 2.5m^{-1} \rightarrow$

$d_i = \dfrac{1}{2.5m^{-1}} = 0.4m$

(C) (4) refraction is responsible for the bending of light in lenses.

Diffraction

Diffraction is the bending of waves around obstacles, or the spreading of waves as they pass through an opening, most apparent when looking at obstacles or wavelengths having a size of the same order of magnitude as the wavelength. Typically, the smaller the obstacle and longer the wavelength, the greater the diffraction. Taken to the extreme, when a wave is blocked by a small enough opening, the wave passing through the opening actually behaves like a point source for a new wave.

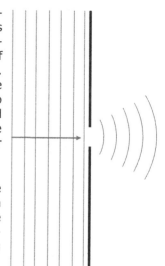

You can observe diffraction quite easily. I'm sure you've heard a noise from a room with an open door even when your ears aren't in a direct line from the sound source. This is a result of diffraction of the sound waves around the door opening (along with some reflection of sound as well).

Thomas Young's Double-Slit Experiment is a famous experiment which utilized diffraction to prove light has properties of waves. Young placed a single-wavelength light source behind a barrier with two narrow slits, allowing only a small portion of the light to pass through each slit. Because the two light waves travel different distances to the screen on which they are projected, you can see effects of both constructive and destructive interference, phenomena that occur only for waves!

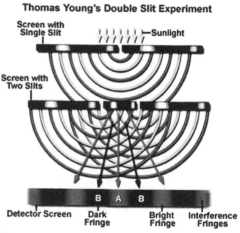

Thomas Young's Double Slit Experiment

Illustration Courtesy of Michael W. Davidson

15.34 Q: Which diagram best represents the shape and direction of a series of wave fronts after they have passed through a small opening in a barrier?

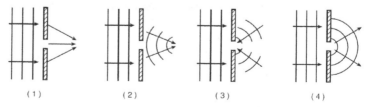

15.34 A: (4) the wave spreads out as it passes through a small opening.

15.35 Q: A beam of monochromatic light approaches a barrier having four openings, A, B, C, and D, of different sizes as shown below.

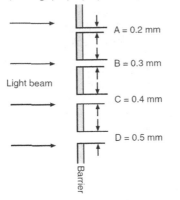

Which opening will cause the greatest diffraction?

15.35 A: (A) has the smallest opening, so will create the most diffraction.

15.36 Q: Parallel wave fronts incident on an opening in a barrier are diffracted. For which combination of wavelength and size of opening will diffraction effects be greatest?

(1) short wavelength and narrow opening

(2) short wavelength and wide opening

(3) long wavelength and narrow opening

(4) long wavelength and wide opening

15.36 A: (3) long wavelength and narrow opening produces the greatest diffraction.

15.37 Q: A wave of constant wavelength diffracts as it passes through an opening in a barrier. As the size of the opening is increased, the diffraction effects

(1) decrease

(2) increase

(3) remain the same

15.37 A: (1) As the size of the opening increases, the amount of diffraction decreases.

15.38 Q: The diagram below shows a plane wave passing through a small opening in a barrier.

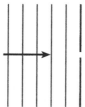

Sketch four wave fronts after they have passed through the barrier.

15.38 A:

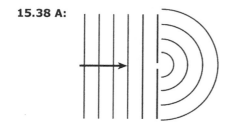

Electromagnetic Spectrum

Unlike mechanical waves, electromagnetic (EM) waves do not require a medium in which to travel. They consist of an electric field component and a magnetic field component oriented perpendicular to each other and to the wave velocity, and are caused by vibrating electrical charges. The orientation of the electric field and magnetic field components of an electromagnetic wave can be visualized below.

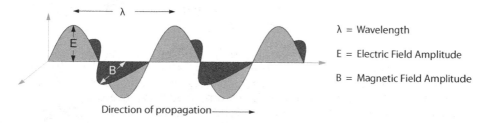

λ = Wavelength

E = Electric Field Amplitude

B = Magnetic Field Amplitude

Direction of propagation

The speed of all electromagnetic waves in a vacuum is approximately 3×10^8 m/s. This constant is so important in physics that it is represented by the letter c, and is, according to our current understanding of the universe, the fastest possible speed anything in the universe can travel.

Since c is a constant for all EM waves in a vacuum, the product of frequency and wavelength must be a constant. Therefore, at higher frequencies, EM waves have a shorter wavelength, and at lower frequencies, EM waves have a longer wavelength. If the EM wave travels into a new medium, its speed can decrease, and because frequency remains constant, its wavelength would also decrease.

It is the frequency of an EM wave that determines its characteristics. The relationship between frequency and wavelength in a vacuum for various types of EM waves is depicted in the Electromagnetic Spectrum. This diagram can be useful for answering questions and solving problems involving electromagnetic waves.

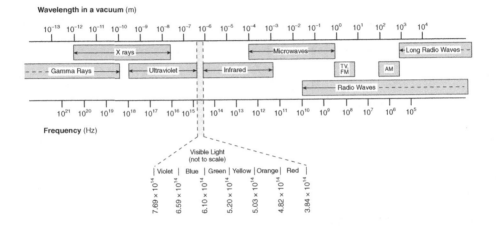

The electromagnetic spectrum describes the types of electromagnetic waves observed at the specified frequencies and wavelength. It is also important to note that the energy of an electromagnetic wave is directly related to its frequency, therefore higher frequency (shorter wavelength) EM waves have more energy than lower frequency (longer wavelength) EM waves.

An x-ray, therefore, has considerably more energy than an AM radio wave! Using the diagram, more energetic waves are shown on the left side of the EM Spectrum, and less energetic waves are shown to the right on the EM Spectrum. You'll explore the energy of EM radiation further in the Modern Physics chapter.

15.39 Q: Which color of light has a wavelength of 5.0×10⁻⁷ meters in air?

(1) blue

(2) green

(3) orange

(4) violet

15.39 A: (2) First find the frequency using v=fλ, then use the Electromagnetic Spectrum to determine the correct color based on the frequency.

$$v = f\lambda$$

$$f = \frac{v}{\lambda} = \frac{c}{\lambda} = \frac{3 \times 10^8 \, ^m/_s}{5.0 \times 10^{-7} \, m} = 6 \times 10^{14} \, Hz$$

15.40 Q: A television remote control is used to direct pulses of electromagnetic radiation to a receiver on a television. This communication from the remote control to the television illustrates that electromagnetic radiation

(1) is a longitudinal wave

(2) possesses energy inversely proportional to its frequency

(3) diffracts and accelerates in air

(4) transfers energy without transferring mass

15.40 A: (4) transfers energy without transferring mass.

15.41 Q: A microwave and an x ray are traveling in a vacuum. Compared to the wavelength and period of the microwave, the x ray has a wavelength that is

(1) longer and a period that is shorter

(2) longer and a period that is longer

(3) shorter and a period that is longer

(4) shorter and a period that is shorter

15.41 A: (4) shorter and a period that is shorter.

15.42 Q: A 1.50×10⁻⁶-meter-long segment of an electromagnetic wave having a frequency of 6×10¹⁴ hertz is represented below.

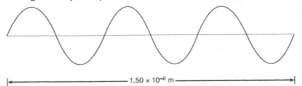

(A) Mark two points on the wave that are in phase with each other. Label each point with the letter P.

(B) Which type of electromagnetic wave does the segment in the diagram represent?

15.42 A: (A)

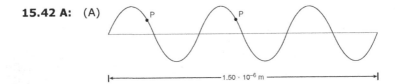

(B) green light (visible light).

15.43 Q: What is the period of a 60-hertz electromagnetic wave traveling at 3.0×10⁸ meters per second?

(1) 1.7×10⁻² s

(2) 2.0×10⁻⁷ s

(3) 6.0×10¹ s

(4) 5.0×10⁶ s

15.43 A: (1) $T = \dfrac{1}{f} = \dfrac{1}{60Hz} = 0.017s$

Chapter 16: Modern Physics

"God does not play dice with the cosmos."
— Albert Einstein

"Einstein, don't tell God what to do."
— Niels Bohr

Objectives

1. Explain the wave-particle duality of light.
2. Calculate the energy of a photon from its wave characteristics.
3. Calculate the energy of an absorbed or emitted photon from an energy level diagram.
4. Explain the quantum nature of atomic energy levels.
5. Explain the Rutherford and Bohr models of the atom.
6. Explain the universal conservation laws.
7. Recognize the fundamental source of all energy in the universe as the conversion of mass into energy.
8. Understand and use the mass-energy equivalence equation.
9. Understand that atomic particles are composed of subnuclear particles.
10. Explain how the nucleus is a conglomeration of quarks which combine to form protons and neutrons.
11. Understand that each elementary particle has a corresponding anti-particle.
12. Utilize Standard Model diagrams to solve basic particle physics problems.
13. Define the known fundamental forces in the universe and rank them in order of relative strength.

Modern Physics refers largely to advancements in physics from the 1900s to the present, extending the models of Newtonian (classical) mechanics and electricity and magnetism to the extremes of the very small, the very large, the very slow and the very fast. Modern Physics can encompass a tremendous variety of topics, which will be explored briefly in this book. Key topics for this exploration include:

- models of the atom
- sub-atomic structure
- universal conservation laws
- mass-energy equivalence
- fundamental forces in the universe
- the dual nature of electromagnetic radiation
- the quantum nature of atomic energy levels

Wave-Particle Duality

Although electromagnetic waves exhibit many characteristics and properties of waves, they can also exhibit some characteristics and properties of particles. These "particles" are called **photons**. Light (and all EM radiation), therefore, has a dual nature. At times, light acts like a wave, and at other times it acts like a particle.

Characteristics of light that indicate light behaves like a wave include:

- Diffraction
- Interference
- Doppler Effect
- Young's Double-Slit Experiment

Characteristics of light that indicate light also acts as a particle include Blackbody Radiation, the Photoelectric Effect, and the Compton Effect.

16.01 Q: Light demonstrates the characteristics of

  (1) particles, only

  (2) waves, only

  (3) both particles and waves

  (4) neither particles nor waves

16.01 A: (3) both particles and waves.

Chapter 16: Modern Physics

16.02 Q: Which phenomenon provides evidence that light has a wave nature?

(1) emission of light from an energy-level transition in a hydrogen atom

(2) diffraction of light passing through a narrow opening

(3) absorption of light by a black sheet of paper

(4) reflection of light from a mirror

16.02 A: (2) diffraction is a phenomenon only applicable to waves.

Blackbody Radiation

The radiation emitted from a very hot object (known as **black-body radiation**) didn't align with physicists' understanding of light as a wave. Specifically, very hot objects emitted radiation in a specific spectrum of frequencies and intensities, which varied with the temperature of the object. Hotter objects had higher intensities at lower wavelengths (toward the blue/UV end of the spectrum), and cooler objects emitted more intensity at higher wavelengths (toward the red/infrared end of the spectrum). Physicists expected that at very short wavelengths the energy radiated would become very large, in contrast to observed spectra. This problem was known as the ultraviolet catastrophe.

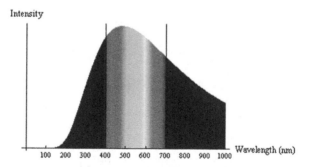

German physicist Max Planck solved this puzzle by proposing that atoms could only absorb or emit radiation in specific, non-continuous amounts, known as quanta. Energy, therefore, is quantized - it only exists in specific discrete amounts. For his work, Planck was awarded the Nobel Prize in Physics in 1918.

Photoelectric Effect

Further evidence that light behaves like a particle was proposed by Albert Einstein in 1905. Scientists had observed that when EM radiation struck a piece of metal, electrons could be emitted (known as **photoelectrons**). What was troubling was that not all EM radiation created photoelectrons. Regardless of what intensity of light was incident upon the metal, the only variable that effected the creation of photoelectrons was the frequency of the light.

If energy exists only in specific, discrete amounts, EM radiation exists in specific discrete amounts, and these smallest possible "pieces" of EM radiation are known as **photons**. A photon has zero mass and zero charge, and because it is a type of EM radiation, its velocity in a vacuum is equal to c $(3\times10^8$ m/s). The energy of each photon of light is therefore quantized and is related to its frequency by the equation:

$$E_{photon} = hf = \frac{hc}{\lambda}$$

In this equation, the value of h, known as **Planck's Constant**, is given as 6.63×10^{-34} J•s.

Einstein proposed that the electrons in the metal object were held in an "energy well," and had to absorb at least enough energy to pull the electron out of the energy well in order to emit a photoelectron. The electrons in the metal would not be released unless they absorbed a single photon with that minimum amount of energy, known as the work function (φ) of the metal. The frequency of this photon is known as the **cutoff frequency** of the metal. Any excess absorbed energy beyond that required to free the electron became kinetic energy for the photoelectron.

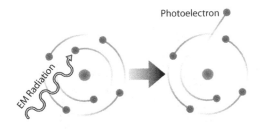

When a high-energy photon of light with energy greater than the energy holding an electron to its nucleus is absorbed by an atom, the electron is emitted as a photoelectron. The kinetic energy of the emitted photoelectron is exactly equal to the amount of energy holding the electron to the nucleus (the work function) subtracted from the energy of the absorbed photon.

$$KE = hf - \phi$$

This theory extended Planck's work and inferred the particle-like behavior of photons of light. Photoelectrons would be ejected from the metal only if they absorbed a photon of light with frequency greater than or equal to a minimum threshold frequency, corresponding to the energy of a photon equal to the metal's "electron well" energy for the most loosely held electrons. Regardless of the intensity of the incident EM radiation, only EM radiation at or above the threshold frequency could produce photoelectrons.

16.03 Q: A photon of light traveling through space with a wavelength of 6×10^{-7} meters has an energy of

(1) 4.0×10^{-40} J
(2) 3.3×10^{-19} J
(3) 5.4×10^{10} J
(4) 5.0×10^{14} J

16.03 A: (2) $E = \dfrac{hc}{\lambda} = \dfrac{(6.63 \times 10^{-34} J \bullet s)(3 \times 10^{8} \, ^{m}/_{s})}{6 \times 10^{-7} m} = 3.3 \times 10^{-19} J$

16.04 Q: The spectrum of visible light emitted during transitions in excited hydrogen atoms is composed of blue, green, red, and violet lines. What characteristic of light determines the amount of energy carried by a photon of that light?

(1) amplitude
(2) frequency
(3) phase
(4) velocity

16.04 A: (2) frequency determines the energy carried by a photon.

16.05 Q: Determine the frequency of a photon whose energy is 3×10^{-19} joule.

16.05 A: $E = hf$

$$f = \frac{E}{h} = \frac{3 \times 10^{-19} J}{6.63 \times 10^{-34} J \bullet s} = 4.5 \times 10^{14} Hz$$

16.06 Q: Light of wavelength 5.0×10^{-7} meter consists of photons having an energy of

(1) 1.1×10^{-48} J
(2) 1.3×10^{-27} J
(3) 4.0×10^{-19} J
(4) 1.7×10^{-5} J

16.06 A: (3) $E = \dfrac{hc}{\lambda} = \dfrac{(6.63 \times 10^{-34} J \bullet s)(3 \times 10^{8} \, ^{m}/_{s})}{5 \times 10^{-7} m} = 4 \times 10^{-19} J$

16.07 Q: A photon has a wavelength of 9×10^{-10} meters. Calculate the energy of this photon in joules.

16.07 A: $E = \dfrac{hc}{\lambda} = \dfrac{(6.63 \times 10^{-34} J \bullet s)(3 \times 10^{8} \frac{m}{s})}{9 \times 10^{-10} m} = 2.2 \times 10^{-16} J$

16.08 Q: The graph below represents the relationship between the energy and the frequency of photons.

Energy vs. Frequency

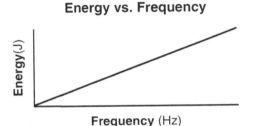

The slope of the graph would be
(1) 6.63×10^{-34} J•s
(2) 6.67×10^{-11} N•m2/kg2
(3) 1.60×10^{-19} J
(4) 1.60×10^{-19} C

16.08 A: (1) The slope of the graph, rise over run, is equivalent to the energy divided by the frequency, which gives you Planck's constant.

16.09 Q: The alpha line in the Balmer series of the hydrogen spectrum consists of light having a wavelength of 6.56×10^{-7} meter.
(A) Calculate the frequency of this light.
(B) Determine the energy in joules of a photon of this light.
(C) Determine the energy in electronvolts of a photon of this light.

16.09 A: (A) $v = f\lambda$

$$f = \frac{v}{\lambda} = \frac{3 \times 10^{8} \frac{m}{s}}{6.56 \times 10^{-7} m} = 4.57 \times 10^{14} Hz$$

(B) $E = \dfrac{hc}{\lambda} = \dfrac{(6.63 \times 10^{-34} J \bullet s)(3 \times 10^{8} \frac{m}{s})}{6.56 \times 10^{-7} m} = 3.03 \times 10^{-19} J$

(C) $3.03 \times 10^{-19} J \times \dfrac{1eV}{1.6 \times 10^{-19} J} = 1.89 eV$

de Broglie Wavelength

Einstein continued to extend his theories around the interaction of photons and atomic particles, going so far as to hypothesize that photons could have momentum, also a particle property, even though they had no mass.

In 1922, American physicist Arthur Compton shot an X-ray photon at a graphite target to observe the collision between the photon and one of the graphite atom's electrons. Compton observed that when the photon collided with an electron, a photoelectron was emitted, but the original X-ray was also scattered and emitted, and with a longer wavelength (indicating it had lost energy).

Further, the longer wavelength also indicated that the photon must have lost momentum. A detailed analysis showed that the energy and momentum lost by the X-ray was exactly equal to the energy and momentum gained by the photoelectron. Compton therefore concluded that not only do photons have momentum, they also obey the laws of conservation of energy and conservation of momentum!

In 1923, French physicist Louis de Broglie took Compton's finding one step further. He stated that if EM waves can behave as moving particles, it would only make sense that a moving particle should exhibit wave properties. De Broglie's hypothesis was confirmed by shooting electrons through a double slit, similar to Young's Double Slit Experiment, and observing a diffraction pattern. The smaller the particle, the more apparent its wave properties are. The wavelength of a moving particle, now known as the **de Broglie Wavelength**, is given by:

$$\lambda = \frac{h}{p}$$

16.10 Q: Moving electrons are found to exhibit properties of
    (1) particles, only
    (2) waves, only
    (3) both particles and waves
    (4) neither particles nor waves

16.10 A: (3) moving particles have both particle and wave properties.

16.11 Q: Which phenomenon best supports the theory that matter has a wave nature?
    (1) electron momentum
    (2) electron diffraction
    (3) photon momentum
    (4) photon diffraction

16.11 A: (2) The diffraction of electrons indicates that electrons behave like waves.

16.12 Q: Wave-particle duality is most apparent in analyzing the motion of
(1) a baseball
(2) a space shuttle
(3) a galaxy
(4) an electron

16.12 A: (4) Wave-particle duality is most easily observed in small objects.

Models of the Atom

In the early 1900s, scientists around the world began to refine and revise our understanding of atomic structure and sub-atomic particles. Scientists understood that matter was made up of atoms, and J.J. Thompson had shown that atoms contained very small negative particles known as electrons, but beyond that, the atom remained a mystery.

New Zealand scientist Ernest Rutherford devised an experiment to better understand the rest of the atom. The experiment, known as **Rutherford's Gold Foil Experiment**, involved shooting alpha particles (helium nuclei) at a very thin sheet of gold foil and observing the deflection of the particles after passing through the gold foil. Rutherford found that although most of the particles went through undeflected, a significant number of alpha particles were deflected by large amounts. Using an analysis based around Coulomb's Law and the conservation of momentum, Rutherford concluded that:

1. Atoms have a small, massive, positive nucleus at the center.
2. Electrons must orbit the nucleus.
3. Most of the atom is made up of empty space.

Rutherford's model was incomplete, though, in that it didn't account for a number of effects predicted by classical physics. Classical physics predicted that if the electron orbits the atom, it is constantly accelerating, and should therefore emit photons of EM radiation. Because the atom emits photons, it should be losing energy, therefore the orbit of the electron would quickly decay into the nucleus and the atom would be unstable. Further, elements were found to emit and absorb EM radiation only at specific frequencies, which did not correlate to Rutherford's theory.

Following Rutherford's discovery, Danish physicist Niels Bohr traveled to England to join Rutherford's research group and refine Rutherford's model of the atom. Instead of focusing on all atoms, Bohr confined his research to developing a model of the simple hydrogen atom. Bohr's model made the following assumptions:

1. Electrons don't lose energy as they accelerate around the nucleus. Instead, energy is quantized. Electrons can only exist at specific discrete energy levels.
2. Each atom allows only a limited number of specific orbits (electrons) at each energy level.
3. To change energy levels, an electron must absorb or emit a photon of energy exactly equal to the difference between the electron's initial and final energy levels: $E_{photon} = E_i - E_f$

Bohr's Model, therefore, was able to explain some of the limitations of Rutherford's Model. Further, Bohr was able to use his model to predict the frequencies of photons emitted and absorbed by hydrogen, explaining Rutherford's problem of emission and absorption spectra! For his work, Bohr was awarded the Nobel Prize in Physics in 1922.

16.13 Q: Calculate the energy and wavelength of the emitted photon when an electron moves from an energy level of -1.51 eV to -13.6 eV.

16.13 A: $E_{photon} = E_i - E_f = (-1.51eV) - (-13.6eV) = 12.09eV$

$$E_{photon} = \frac{hc}{\lambda}$$

$$\lambda = \frac{hc}{E_{photon}} = \frac{(6.63 \times 10^{-34} J \bullet s)(3 \times 10^8 \text{ }^m/_s)}{(12.09eV)(1.6 \times 10^{-19} \text{ }^J/_{eV})} = 1.03 \times 10^{-7} m$$

Energy Level Diagrams

A useful tool for visualizing the allowed energy levels in an atom is the energy level diagram. Two of these diagrams (one for hydrogen and one for mercury) are provided for you on the following page. In each of these diagrams, the n=1 energy state is the lowest possible energy for an electron of that atom, known as the ground state. The energy corresponding to n=1 is shown on the right side of the diagram in electronvolts. So, for hydrogen, the ground state is a level of -13.6 eV.

Hydrogen

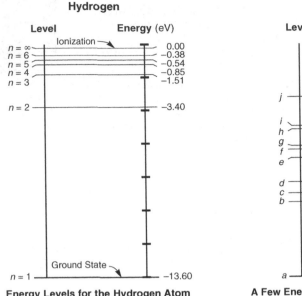

Energy Levels for the Hydrogen Atom

Mercury

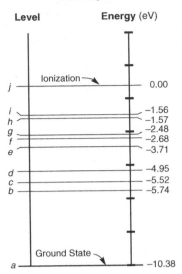

A Few Energy Levels for the Mercury Atom

The energy levels are negative to indicate that the electron is bound by the nucleus of the atom. If the electron reaches 0 eV, it is no longer bound by the atom and can be emitted as a photoelectron (i.e. the atom becomes ionized). Any remaining energy becomes the kinetic energy of the photoelectron.

16.14 Q: An electron in a hydrogen atom drops from the n=3 to the n=2 state. Determine the energy of the emitted radiation.

16.14 A:

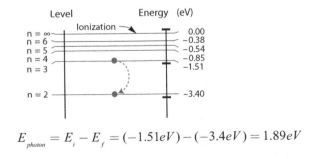

$$E_{photon} = E_i - E_f = (-1.51eV) - (-3.4eV) = 1.89eV$$

16.15 Q: Which type of photon is emitted when an electron in a hydrogen atom drops from the n = 2 to the n = 1 energy level?

(1) ultraviolet

(2) visible light

(3) infrared

(4) radio wave

16.15 A: (1) First find the amount of energy emitted in electron volts, convert that energy to Joules, then find the frequency of the emitted radiation, which you can look up on the EM Spectrum to determine the radiation type.

$$E_{photon} = E_i - E_f = (-3.4eV) - (-13.6eV) = 10.2eV$$

$$10.2eV \times \frac{1.6 \times 10^{-19} J}{1eV} = 1.63 \times 10^{-18} J$$

$$E = hf \quad f = \frac{E}{h} = \frac{1.63 \times 10^{-18} J}{6.63 \times 10^{-34} J \bullet s} = 2.46 \times 10^{15} Hz$$

16.16 Q: Base your answers on the Energy Level Diagram for Hydrogen on the previous page.

(A) Determine the energy, in electronvolts, of a photon emitted by an electron as it moves from the n = 6 to the n = 2 energy level in a hydrogen atom.

(B) Convert the energy of the photon to joules.

(C) Calculate the frequency of the emitted photon.

(D) Is this the only energy and/or frequency that an electron in the n = 6 energy level of a hydrogen atom could emit? Explain your answer.

16.16 A: (A) $E_{photon} = E_i - E_f = (-0.38eV) - (-3.4eV) = 3.02eV$

(B) $3.02eV \times \dfrac{1.6 \times 10^{-19} J}{1eV} = 4.83 \times 10^{-19} J$

(C) $E = hf \rightarrow f = \dfrac{E}{h} = \dfrac{4.83 \times 10^{-19} J}{6.63 \times 10^{-34} J \bullet s} = 7.29 \times 10^{14} Hz$

(D) No, this is not the only energy and/or frequency that an electron in the n=6 energy level of a hydrogen atom could emit. The electron can return to any of the five lower energy levels.

16.17 Q: An electron in a mercury atom drops from energy level f to energy level c by emitting a photon having an energy of

(1) 8.20 eV

(2) 5.52 eV

(3) 2.84 eV

(4) 2.68 eV

16.17 A: (3) $E_{photon} = E_i - E_f = (-2.68eV) - (-5.52eV) = 2.84eV$

16.18 Q: A mercury atom in the ground state absorbs 20 electronvolts of energy and is ionized by losing an electron. How much kinetic energy does this electron have after the ionization?

(1) 6.40 eV

(2) 9.62 eV

(3) 10.38 eV

(4) 13.60 eV

16.18 A: (2) The ionization energy for an electron in the ground state of a mercury atom is 10.38 eV according to the Mercury Energy Level Diagram. If the atom absorbs 20 eV of energy, and uses up 10.38 eV in ionizing the electron, the electron has a leftover energy of 9.62 eV, which must be the electron's kinetic energy.

16.19 Q: A hydrogen atom with an electron initially in the n = 2 level is excited further until the electron is in the n = 4 level. This energy level change occurs because the atom has

(1) absorbed a 0.85-eV photon

(2) emitted a 0.85-eV photon

(3) absorbed a 2.55-eV photon

(4) emitted a 2.55-eV photon

16.19 A: (3) absorbed a 2.55-eV photon.

Atomic Spectra

Once you understand the energy level diagram, it quickly becomes obvious that atoms can only emit certain frequencies of photons, correlating to the difference between energy levels as an electron falls from a higher energy state to a lower energy state. In similar fashion, electrons can only absorb photons with energy equal to the difference in energy levels as the electron jumps from a lower to a higher energy state. This leads to unique atomic spectra of emitted radiation for each element.

An object that is heated to the point where it glows (**incandescence**) emits a continuous energy spectrum, described as blackbody radiation.

If a gas-discharge lamp is made from mercury vapor, the mercury vapor is made to emit light by application of a high electrical potential. The light emitted by the mercury vapor is created by electrons in higher energy states falling to lower energy states, therefore the photons emitted correspond directly in wavelength to the difference in energy levels of the electrons. This creates a unique spectrum of frequencies which can be observed by separating the colors using a prism, known as an emission spectrum. By analyzing the emission spectra of various objects, scientists can determine the composition of those objects.

In similar fashion, if light of all colors is shone through a cold gas, the gas will only absorb the frequencies corresponding to photon energies exactly equal to the difference between the gas's atomic energy levels. This creates a spectrum with all colors except those absorbed by the gas, known as an absorption spectrum.

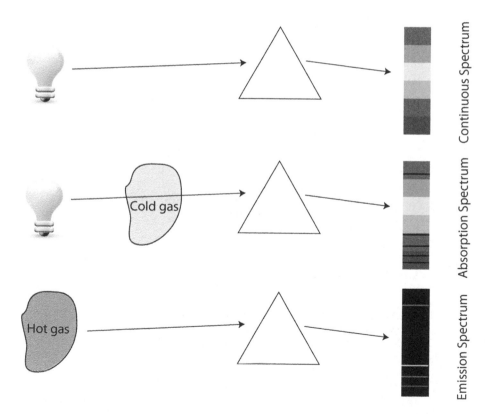

16.20 Q: The bright-line emission spectrum of an element can best be explained by

(1) electrons transitioning between discrete energy levels in the atoms of that element

(2) protons acting as both particles and waves

(3) electrons being located in the nucleus

(4) protons being dispersed uniformly throughout the atoms of that element

16.20 A: (1) bright-line emission spectra are created by electrons moving between energy levels, giving off photons of energy equal to the difference in energy levels.

16.21 Q: The diagram below represents the bright-line spectra of four elements, A, B, C, and D, and the spectrum of an unknown gaseous sample.

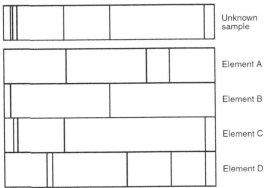

Based on comparisons of these spectra, which two elements are found in the unknown sample?

(1) A and B

(2) A and D

(3) B and C

(4) C and D

16.21 A: (3) Elements B and C have bright lines corresponding to the unknown sample.

Mass—Energy Equivalence

In 1905, in a paper titled "Does the Inertia of a Body Depend Upon Its Energy Content," Albert Einstein proposed the revolutionary concept that an object's mass is a measure of how much energy that object contains, opening a door to a host of world-changing developments, eventually leading us to the major understanding that the source of all energy in the universe is, ultimately, the conversion of mass into energy!

If mass is a measure of an object's energy, we need to re-evaluate our statements of the law of conservation of mass and the law of conservation of energy. Up to this point, we have thought of these as separate statements of fact in the universe. Based on Einstein's discovery, however, mass and energy are two concepts effectively describing the same thing, therefore we could more appropriately combine these two laws into a single law: the law of conservation of mass-energy. This law states that mass-energy cannot be created nor destroyed.

The concept of mass-energy is one that is often misunderstood and oftentimes argued in terms of semantics. For example, a popular argument states that the concept of mass-energy equivalence means that mass can be converted to energy, and energy can be converted to mass. Many would disagree that this can occur, countering that since mass and energy are effectively the same thing, you can't convert one to the other. For our purposes, we'll save these arguments for future courses of study. Instead, we will focus on a basic conceptual understanding.

The universal conservation laws studied so far in this book course include:

- Conservation of Mass-Energy
- Conservation of Charge
- Conservation of Momentum

Einstein's famous formula, $E=mc^2$, relates the amount of energy contained in matter to the mass times the speed of light in a vacuum ($c=3\times10^8$ m/s) squared.

Theoretically, you could determine the amount of energy represented by 1 kilogram of matter as follows:

16.22 Q: What is the energy equivalent of 1 kilogram of matter?

16.22 A: $E = mc^2 = (1kg)(3 \times 10^8 \, ^m\!/_s)^2 = 9 \times 10^{16} \, J$

This is a very large amount of energy. To put it in perspective, the energy equivalent of a large pickup truck is in the same order of magnitude of the total annual energy consumption of the United States!

More practically, however, it is not realistic to convert large quantities of mass completely into energy. Current practice revolves around converting small amounts of mass into energy in nuclear processes. Typically these masses are so small that measuring in units of kilograms is cumbersome. Instead, scientists often work with the much smaller **universal mass unit** (u), which is equal in mass to one-twelfth the mass of a single atom of Carbon-12. The mass of a proton and neutron, therefore, is close to 1u, and the mass of an electron is close to 5×10^{-4}u. In precise terms, 1u=1.66053886×10^{-27}kg.

One universal mass unit (1u) completely converted to energy is equivalent to 931 MeV. Because mass and energy are different forms of the same thing, this could even be considered a unit conversion problem.

16.23 Q: If a deuterium nucleus has a mass of 1.53×10^{-3} universal mass units (u) less than its components, how much energy does its mass represent?

16.23 A: $(1.53 \times 10^{-3} u) \times \dfrac{9.31 \times 10^2 \, MeV}{1u} = 1.42 \, MeV$

The nucleus of an atom consists of positively charged protons and neutral neutrons. Collectively, these nuclear particles are known as nucleons. Protons repel each other electrically, so why doesn't the nucleus fly apart? There is another force which holds nucleons together, known as the **strong nuclear force**. This extremely strong force overcomes the electrical repulsion of the protons, but it is only effective over very small distances.

Chapter 16: Modern Physics

Because nucleons are held together by the strong nuclear force, you must add energy to the system to break apart the nucleus. The energy required to break apart the nucleus is known as the **binding energy** of the nucleus.

If measured carefully, we find that the mass of a stable nucleus is actually slightly less than the mass of its individual component nucleons. The difference in mass between the entire nucleus and the sum of its component parts is known as the **mass defect** (Δm). The binding energy of the nucleus, therefore, must be the energy equivalent of the mass defect due to the law of conservation of mass-energy: $E_{binding} = \Delta mc^2$.

Fission is the process in which a nucleus splits into two or more nuclei. For heavy (larger) nuclei such as Uranium-235, the mass of the original nucleus is greater than the sum of the mass of the fission products. Where did this mass go? It is released as energy! A commonly used fission reaction involves shooting a neutron at an atom of Uranium-235, which briefly becomes Uranium-236, an unstable isotope. The Uranium-236 atom then fissions into a Barium-141 atom and a Krypton-92 atom, releasing its excess energy while also sending out three more neutrons to continue a chain reaction! This process is responsible for our nuclear power plants, and is also the basis (in an uncontrolled reaction) of atomic fission bombs.

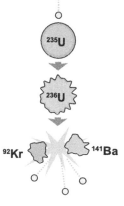

Fusion, on the other hand, is the process of combining two or more smaller nuclei into a larger nucleus. If this occurs with small nuclei, the product of the reaction may have a smaller mass its precursors, thereby releasing energy as part of the reaction. This is the basic nuclear reaction that fuels our sun and the stars as hydrogen atoms combine to form helium. This is also the basis of atomic hydrogen bombs.

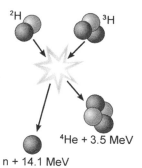

Nuclear fusion holds tremendous potential as a clean source of power with widely available source material (we can create hydrogen from water). The most promising fusion reaction for controlled energy production fuses two isotopes of hydrogen known as deuterium and tritium to form a helium nucleus and a neutron, as well as an extra neutron, while releasing a considerable amount of energy. Currently, creating a sustainable, controlled fusion reaction that outputs more energy than is required to start the reaction has not yet been demonstrated, but remains an area of focus for scientists and engineers.

16.24 Q: In the first nuclear reaction using a particle accelerator, accelerated protons bombarded lithium atoms, producing alpha particles and energy. The energy resulted from the conversion of mass into energy. The reaction can be written as shown below.

$$_1^1H + _3^7Li \rightarrow _2^4He + energy$$

Data Table

Particle	Symbol	Mass (u)
proton	$_1^1H$	1.007 83
lithium atom	$_3^7Li$	7.016 00
alpha particle	$_2^4He$	4.002 60

(A) Determine the difference between the total mass of a proton plus a lithium atom, and the total mass of two alpha particles, in universal mass units.

(B) Determine the energy in megaelectronvolts produced in the reaction of a proton with a lithium atom.

16.24 A: (A) $(1.00783u + 7.01600u) - 2(4.00260u) = 0.01863u$

(B) $0.01863u \times \dfrac{9.31 \times 10^2\ MeV}{1u} = 17.3\ MeV$

16.25 Q: The energy produced by the complete conversion of 2×10^{-5} kilograms of mass into energy is

(1) 1.8 TJ

(2) 6.0 GJ

(3) 1.8 MJ

(4) 6.0 kJ

16.25 A: (1) $E = mc^2 = (2 \times 10^{-5} kg)(3 \times 10^8\ ^m\!/_s)^2 = 1.8 \times 10^{12} J = 1.8 TJ$

16.26 Q: A tritium nucleus is formed by combining two neutrons and a proton. The mass of this nucleus is 9.106×10^{-3} universal mass unit less than the combined mass of the particles from which it is formed. Approximately how much energy is released when this nucleus is formed?

(1) 8.48×10^{-2} MeV

(2) 2.73 MeV

(3) 8.48 MeV

(4) 273 MeV

16.26 A: (3) $9.106 \times 10^{-3} u \times \dfrac{9.31 \times 10^{2}\ MeV}{1u} = 8.48\ MeV$

16.27 Q: The energy equivalent of 5×10^{-3} kilogram is

(1) 8.0×10^{5} J

(2) 1.5×10^{6} J

(3) 4.5×10^{14} J

(4) 3.0×10^{19} J

16.27 A: (3) $E = mc^{2} = (5 \times 10^{-3} kg)(3 \times 10^{8}\ ^{m}\!/_{s})^{2} = 4.5 \times 10^{14}\ J$

16.28 Q: After a uranium nucleus emits an alpha particle, the total mass of the new nucleus and the alpha particle is less than the mass of the original uranium nucleus. Explain what happens to the missing mass.

16.28 A: The missing mass is converted into energy.

The Standard Model

As you've learned previously, the atom is the smallest part of an element (such as oxygen) that has the characteristics of the element. Atoms are made up of very small negatively charged electrons surrounding the much larger nucleus. The nucleus is composed of positively charged protons and neutral neutrons. The positively charged protons exert a repelling electrical force upon each other, but the strong nuclear force holds the protons and neutrons together in the nucleus.

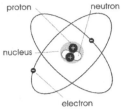

This completely summarized scientists' understanding of atomic structure until the 1930s, when scientists began to discover evidence that there was more to the picture and that protons and nucleons were made up of even smaller particles. This launched the particle physics movement, which, to this day, continues to challenge the understanding of the entire universe by exploring the structure of the atom.

In addition to standard matter, researchers have discovered the existence of antimatter. **Antimatter** is matter made up of particles with the same mass as regular matter particles, but opposite charges and other characteristics. An **antiproton** is a particle with the same mass as a proton, but a negative (opposite) charge. A **positron** has the same mass as an electron, but a positive charge. An **antineutron** has the same mass as a neutron, but has other characteristics opposite that of the neutron.

When a matter particle and its corresponding antimatter particle meet, the particles may combine to **annihilate** each other, resulting in the complete conversion of both particles into energy consistent with the mass-energy equivalence equation: $E=mc^2$.

16.29 Q: A proton and an antiproton collide and completely annihilate each other. How much energy is released? ($m_{proton}=1.67\times10^{-27}$kg)

16.29 A: $E = mc^2 = 2(1.67\times10^{-27}\,kg)(3\times10^8\,\text{m/s})^2 = 3\times10^{-10}\,J$

This book has dealt with many types of forces, ranging from contact forces such as tensions and normal forces to field forces such as the electrical force and gravitational force. When observed from their most basic aspects, however, all observed forces in the universe can be consolidated into four fundamental forces. They are, from strongest to weakest:

1. Strong Nuclear Force: holds protons and neutrons together in the nucleus
2. Electromagnetic Force: electrical and magnetic attraction and

repulsion
3. Weak force: responsible for radioactive beta decay
4. Gravitational Force: attractive force between objects with mass

Understanding these forces remains a topic of scientific research, with current work exploring the possibility that forces are actually conveyed by an exchange of force-carrying particles such as photons, bosons, gluons, and gravitons.

16.30 Q: The particles in a nucleus are held together primarily by the
(1) strong force
(2) gravitational force
(3) electrostatic force
(4) magnetic force

16.30 A: (1) the strong nuclear force holds protons and neutrons together in the nucleus.

16.31 Q: Which fundamental force is primarily responsible for the attraction between protons and electrons?
(1) strong
(2) weak
(3) gravitational
(4) electromagnetic

16.31 A: (4) the electromagnetic force is responsible for the electrostatic attraction and repulsion of charged particles.

16.32 Q: Which statement is true of the strong nuclear force?
(1) It acts over very great distances.
(2) It holds protons and neutrons together.
(3) It is much weaker than gravitational forces.
(4) It repels neutral charges.

16.32 A: (2) The strong nuclear force holds protons and neutrons together.

16.33 Q: The strong force is the force of

(1) repulsion between protons

(2) attraction between protons and electrons

(3) repulsion between nucleons

(4) attraction between nucleons

16.33 A: (4) attraction between nucleons (nucleons are particles in the nucleus such as protons and neutrons).

The current model of sub-atomic structure used to understand matter is known as the Standard Model. Development of this model began in the late 1960s, and has continued through today with contributions from many scientists across the world. The Standard Model explains the interactions of the strong (nuclear), electromagnetic, and weak forces, but has yet to account for the gravitational force. The search for the theorized Higgs Boson is an attempt to better unify and strengthen the Standard Model.

Although the Standard Model itself is a very complicated theory, the basic structure of the model is fairly straightforward. According to the model, all matter is divided into two categories, known as **hadrons** and the much smaller **leptons**. All of the fundamental forces act on hadrons, which include particles such as protons and neutrons. In contrast, the strong nuclear force doesn't act on leptons, so only three fundamental forces act on leptons such as electrons, positrons, muons, tau particles and neutrinos.

Hadrons are further divided into **baryons** and **mesons**. Baryons such as protons and neutrons are composed of three smaller particles known as **quarks**. Charges of baryons are always whole numbers. Mesons are composed of a quark and an anti-quark (for example, an up quark and an anti-down quark).

Classification of Matter

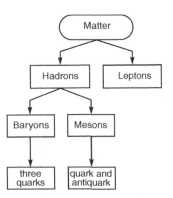

Particles of the Standard Model

Quarks

Name	up	charm	top
Symbol	u	c	t
Charge	$+\frac{2}{3}e$	$+\frac{2}{3}e$	$+\frac{2}{3}e$

	down	strange	bottom
	d	s	b
	$-\frac{1}{3}e$	$-\frac{1}{3}e$	$-\frac{1}{3}e$

Leptons

electron	muon	tau
e	μ	τ
$-1e$	$-1e$	$-1e$

electron neutrino	muon neutrino	tau neutrino
ν_e	ν_μ	ν_τ
0	0	0

Note: For each particle, there is a corresponding antiparticle with a charge opposite that of its associated particle.

Chapter 16: Modern Physics

Scientists have identified six types of quarks. For each of the six types of quarks, there also exists a corresponding anti-quark with an opposite charge. The quarks have rather interesting names: up quark, down quark, charm quark, strange quark, top quark, and bottom quark. Charges on each quark are either one third of an elementary charge, or two thirds of an elementary charge, positive or negative, and the quarks are symbolized by the first letter of their name. For the associated anti-quark, the symbol is the first letter of the anti-quark's name, with a line over the name. For example, the symbol for the up quark is u. The symbol for the anti-up quark is ū.

Similarly, scientists have identified six types of leptons: the electron, the muon, the tau particle, and the electron neutrino, muon neutrino, and tau neutrino. Again, for each of these leptons there also exists an associated anti-lepton. The most familiar lepton, the electron, has a charge of -1e. Its anti-particle, the positron, has a charge of +1e.

Since a proton is made up of three quarks, and has a positive charge, the sum of the charges on its constituent quarks must be equal to one elementary charge. A proton is actually comprised of two up quarks and one down quark. You can verify this by adding up the charges of the proton's constituent quarks (uud).

$$\left(+\frac{2}{3}e\right)+\left(+\frac{2}{3}e\right)+\left(-\frac{1}{3}e\right)=+1e$$

16.34 Q: A neutron is composed of up and down quarks. How many of each type of quark are needed to make a neutron?

16.34 A: The charge on the neutron must sum to zero, and the neutron is a baryon, so it is made up of three quarks. To achieve a total charge of zero, the neutron must be made up of one up quark (+2/3e) and two down quarks (-1/3e).

If the charge on a quark (such as the up quark) is (+2/3)e, the charge of the anti-quark (ū) is (-2/3)e. The anti-quark is the same type of particle, with the same mass, but with the opposite charge.

16.35 Q: What is the charge of the down anti-quark?

16.35 A: The down quark's charge is -1/3e, so the anti-down quark's charge must be +1/3e.

16.36 Q: Compared to the mass and charge of a proton, an antiproton has

    (1) the same mass and the same charge

    (2) greater mass and the same charge

    (3) the same mass and the opposite charge

    (4) greater mass and the opposite charge

16.36 A: (3) the same mass and the opposite charge.

16.37 Q: The diagram below represents the sequence of events (steps 1 through 10) resulting in the production of a D⁻ meson and a D⁺ meson. An electron and a positron (antielectron) collide (step 1), annihilate each other (step 2), and become energy (step 3). This energy produces an anticharm quark and a charm quark (step 4), which then split apart (steps 5 through 7). As they split, a down quark and an antidown quark are formed, leading to the final production of a D⁻ meson and a D⁺ meson (steps 8 through 10).

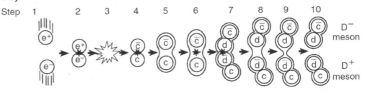

Adapted from: Electon/Positron Annihilation http:/www.particleadventure.org/frameless/eedd.html 7/23/2007

Which statement best describes the changes that occur in this sequence of events?

(1) Energy is converted into matter and then matter is converted into energy.

(2) Matter is converted into energy and then energy is converted into matter.

(3) Isolated quarks are being formed from baryons.

(4) Hadrons are being converted into leptons.

16.37 A: (2) Particles are converted into energy, which is then converted into particles.

16.38 Q: What fundamental force holds quarks together to form particles such as protons and neutrons?

    (1) electromagnetic force

    (2) gravitational force

    (3) strong force

    (4) weak force

16.38 A: (3) the strong force holds particles together in the nucleus.

16.39 Q: A particle unaffected by an electric field could have a quark composition of

(1) css

(2) bbb

(3) udc

(4) uud

16.39 A: (1) In order to not be affected by an electric field, the particle must be neutral. The only combination of quarks which results in a net charge of zero is css.

16.40 Q: For years, theoretical physicists have been refining a mathematical method called lattice quantum chromodynamics to enable them to predict the masses of particles consisting of various combinations of quarks and antiquarks. They recently used the theory to calculate the mass of the rare B_c particle, consisting of a charm quark and a bottom antiquark. The predicted mass of the B_c particle was about six times the mass of a proton.

Shortly after the prediction was made, physicists working at the Fermi National Accelerator Laboratory, Fermilab, were able to measure the mass of the B_c particle experimentally and found it to agree with the theoretical prediction to within a few tenths of a percent. In the experiment, the physicists sent beams of protons and antiprotons moving at 99.999% the speed of light in opposite directions around a ring 1.0 kilometer in radius. The protons and antiprotons were kept in their circular paths by powerful electromagnets. When the protons and antiprotons collided, their energy produced numerous new particles, including the elusive B_c.

These results indicate that lattice quantum chromodynamics is a powerful tool not only for confirming the masses of existing particles, but also for predicting the masses of particles that have yet to be discovered in the laboratory.

(A) Identify the class of matter to which the B_c particle belongs.

(B) Determine both the sign and the magnitude of the charge of the B_c particle in elementary charges.

(C) Explain how it is possible for a colliding proton and antiproton to produce a particle with six times the mass of either.

16.40 A: (A) Meson or Hadron

(B) +1e. The charge on a charm quark (+2/3e) and a bottom antiquark (+1/3e) sum to +1e.

(C) Energy is converted to mass.

16.41 Q: The charge of an antistrange quark is approximately

    (1) $+5.33 \times 10^{-20}$ C

    (2) -5.33×10^{-20} C

    (3) $+5.33 \times 10^{20}$ C

    (4) -5.33×10^{20} C

16.41 A: (1) charge on an antistrange quark is $+1/3e = +5.33 \times 10^{-20}$ C.

Einstein's Relativity

Albert Einstein was puzzled by the concept of relative motion, especially when applied to situations of electromagnetic and gravitational forces. Einstein postulated that there are no instantaneous reactions in nature, therefore there must be a maximum possible speed for any reaction, which is the speed of light in a vacuum (c). Further, the speed of light in the vacuum must be the same for all observers, whether moving or at rest.

Einstein felt confident of these postulates, but soon realized that the repercussions of these postulates would upset the foundations of the known physical world. Eventually, he realized that the only way his postulates could be true in all situations involved a re-imagining of the concept of time, detailed in his proposal of Special Relativity. To begin with, events that appeared to occur simultaneously in one frame of reference did not necessarily occur simultaneously in another frame of reference. The entire concept of simultaneous events was relative!

More specifically, observers moving at different speeds would experience different time intervals. Imagine a modern training traveling at high speed. A laser, in the exact center of the train, turns on and shoots a laser beam toward the front and the back of the train. An observer in the center of the train car would see the laser beams would hit the front and back windows of the train simultaneously. An observer on a lawn chair outside the train watching the train pass by, however, would see the laser exit the back window of the train first, since the back of the train is moving to meet the laser beam, while the front of the train is moving away from the laser beam. The observer outside the train does not see simultaneous events, while the observer inside the train does see simultaneous events. Time is therefore relative to the observer.

The implications of these findings are wide-spread and complex. Objects traveling at high velocities relative to an observer experience what is known as time dilation. What is experienced as a short time interval by the high-speed object is experienced as a longer time interval by the observer.

         Chapter 16: Modern Physics

A famous thought experiment involves two identical twins on Earth. Suppose one twin leaves Earth at the age of 20 in an imaginary spaceship and travels at 90% of the speed of light (0.9c) a distance of 10 light years, then turns around and returns to Earth. The second twin remains on Earth. The space-faring twin returns having experienced a trip of just under 20 years time (under 20 years due to some other secondary relativistic effects), and has aged 20 years, returning as a 40-year-old, while the twin who remained on Earth experienced roughly 44 years while the sibling was in space, and is now 64 years old.

Further, as objects travel at higher speeds, their length contracts compared to the stationary observer. Further yet, as objects move faster and faster, it takes more and more energy to accelerate them further, therefore mass can never be accelerated to the speed of light.

Einstein generalized his work in the Theory of General Relativity, where he proposed that space and time are intertwined in a universal fabric known as spacetime. Large masses have the ability to bend the fabric of spacetime, leading to what you experience as gravity.

If all this sounds a bit complex and confusing, you're not alone. Entire courses and careers have been devoted to exploring and debating these theories. Thankfully, it is easy to find more resources on relativity, both on the web and in print form, written in a variety of formats from illustrated comics to complex mathematical proofs.

Most importantly, these developments serve as a great reminder that despite how much scientists think they know about the universe, there is much more that is yet unknown, and therefore so many more explorations to be undertaken and discoveries to be made that physics continues to grow and evolve on a daily basis.

Index

Index

Index

guitar 312

H

hadrons 364
half-adder 285–286
heat 193
Hertz 113, 304
Higgs Boson 364
hole 273
Hooke's Law 152–172, 162
hydraulics 181
hydrogen 351

I

ideal gas law 200
impulse 90–91
Impulse-Momentum Theorem 91–93, 92
inclines 78–82
index of refraction 327–332, 330
induction 216
inertia 60–62, 139. See also mass
insulators 214
integrated circuit 272
interference. See waves: interference
internal energy 164, 167
inverse square law 118–121, 218–220, 225, 226
Io 120
ion implantation 294
isobaric 205
isochoric 205
isothermal 206
isotherms 206

J

joule 149, 158, 227
Jupiter 120

K

KCL. See Kirchhoff's Current Law
Kelvins 189
Kepler's Laws of Planetary Motion 126–127
Kilby, Jack 272
kilogram 7–8, 225
kinematic equations 34
kinematics 18
Kirchhoff, Gustav 246
Kirchhoff's Current Law 246
Kirchhoff's Voltage Law 246
KVL. See Kirchhoff's Voltage Law

L

law of conservation of energy. See energy: con-
    servation of

law of inertia 67. See Newton's 1st Law of Motion
law of refraction 330
lensmaker's equation 323, 334
leptons 364, 365
lever arm 138
lift 185
light
    speed of 305, 327, 357
line of action 138
lithography 288–289
loudness 307
lucite 331

M

magnetic induction 269
magnetism 262–270
magnets
    flux lines 262
    north pole 262–264
    south pole 262–264
magnification 321
magnitude 11–12, 19
Mars 123
mass 2–3, 61, 64, 123, 139, 357. See also inertia
mass analyzer 294
mass defect 359
mass-energy equivalence 4, 357–361, 362
matter 2, 2–3, 212
medium 298
mercury 351
mesons 364
metals 214, 345
meter 7–8, 19
metric system 7–9
microwaves 298, 341
mirror equation 323
mirrors 320
mks system. See metric system
modern physics 344–376
moles 200
moment arm. See lever arm
moment of inertia 139
momentum 87–104, 349
momentum table 94–98

N

Nernst's Theorem 209
net force 60–61, 62, 64
neutrons 212
Newton, Isaac 59–86, 124
Newtons 62
Newton's 1st Law of Motion 60–62, 67
Newton's 2nd Law for Circular Motion 110
Newton's 2nd Law of Motion 64–67, 74, 78, 110
Newton's 3rd Law of Motion 70–72, 94, 148

Index

Index

Designed to help high school physics students succeed, APlusPhysics is a free online physics resource that focuses on problem solving, understanding, and real-world applications in the context of introductory physics courses such as NY Regents Physics, Honors Physics, and AP Physics B and C courses.

We cover key topics of introductory physics courses, and demonstrate how the principles and applications of high school physics extend outside the classroom through:

- **Courses** - physics tutorials targeted toward specific high school physics courses such as NY Regents Physics, Honors Physics, and AP Physics with sample problems, videos, course notes, and interactive tests. More detailed instruction can be obtained through the APlusPhysics: Regents Physics Essentials book.

- **Projects** – hands-on applications of physics principles to build skills, understanding and confidence.

- **Forums** – an integrated community forum providing students and instructors a haven to discuss and debate current topics in physics; seek out and provide help with challenging high school physics problems; build a better understanding of key concepts; and explore current research and frontiers in physics.

- **Podcasts** - by students, for students, the Physics In Action podcast interviews best-selling authors, explores key concepts, and demonstrates real life implementation of physics principles.

- **Educators** - resources and activities for physics teachers including forensic labs, WebQuests, an educator-only discussion forum and the Semiconductor Technology Enrichment Program.

- **Blogs** - individual and group blogs for students and instructors, providing further opportunities to reflect on concepts and applications while fostering our physics learning community, as well as the Physics In Flux blog which explores redefining the traditional high school physics classroom.

APlusPhysics: Your Guide To
Regents Physics Essentials

Dan Fullerton

APlusPhysics: Your Guide To Regents Physics Essentials is a clear and concise roadmap to the entire New York State Regents Physics curriculum, preparing students for success in their high school physics classes as well as review for high scores on the Regents Physics Exam.

Topics covered include:
- Vectors and Scalars
- Kinematics
- Dynamics
- Circular Motion
- Gravity
- Impulse and Momentum
- Work, Energy, and Power
- Electricity and Magnetism
- Waves and Optics
- Modern Physics

Featuring more than 500 questions from past Regents exams with worked out solutions and detailed illustrations, this book is integrated with the APlusPhysics.com website, which includes online question and answer forums, videos, animations, and supplemental problems to help you master Regents Physics Essentials.

ISBN: 0983563306   ISBN-13: 9780983563303

Made in the USA
Monee, IL
23 March 2022